I0013767

End User License Agreement

This End User License Agreement ("EULA") is a legal agreement between you (either an individual or a single entity) and THE OPEN SERVICE MANAGEMENT ALLIANCE, relating to OSMA courseware, which may include printed or electronic documentation. BY USING THIS COURSEWARE YOU AGREE TO BE BOUND BY THE TERMS OF THIS EULA. IF YOU DO NOT AGREE TO THE TERMS OF THIS EULA, DO NOT USE THIS COURSEWARE.

1. Title. THE OPEN SERVICE MANAGEMENT ALLIANCE, LTD. ("OSMA") is the exclusive owner of the OSMA courseware and related documentation. OSMA is licensing its courseware and documentation to you on a non-exclusive basis as set forth in this Agreement. This license is not a sale. Title to the OSMA courseware, or any copy, modification or merged portion of the OSMA courseware, as well as all related documentation, remains with OSMA always. The courseware and related documentation is protected by copyright laws and international copyright treaties, as well as other intellectual property laws and treaties.

2. Use. OSMA grants to you a non-exclusive, non-transferable license to use the courseware and related documentation in conjunction with related OSMA classes and for personal study.

3. Modification. Modification of any kind of the courseware and related documentation is prohibited.

4. Reproduction. Reproduction of the courseware and related documentation in any form is prohibited.

5. Redistribution and Resale. Redistribution or resale of the courseware and related documentation is prohibited.

6. Limitation of Reverse Engineering. You may not reverse engineer, decompile, or disassemble the courseware and related documentation, except and only to the extent that such activity is expressly permitted by applicable law notwithstanding this limitation.

7. Term and Termination. This license is effective for ten (10) years from time of receipt of the courseware. This license will terminate immediately without notice if you fail to comply with any provision of this license. Once the term ends, this Agreement will automatically terminate as to the licensed courseware item and you may no longer use the courseware and related documentation.

8. Disclaimer of Warranty. YOU EXPRESSLY ACKNOWLEDGE AND AGREE THAT YOUR USE OF THE OSMA COURSEWARE IS AT YOUR SOLE RISK. THE COURSEWARE AND RELATED DOCUMENTATION, IF ANY, ARE PROVIDED "AS IS" AND WITHOUT WARRANTY OF ANY KIND. OSMA EXPRESSLY DISCLAIMS ALL WARRANTIES, EXPRESS OR IMPLIED, INCLUDING, BUT NOT LIMITED TO, THE IMPLIED WARRANTIES OF MERCHANTABILITY, FITNESS FOR A PURPOSE AND NON- INFRINGEMENT. OSMA DOES NOT WARRANT THAT THE INFORMATION CONTAINED IN THE COURSEWARE WILL MEET YOUR REQUIREMENTS, OR THAT THE COURSEWARE WILL BE ERROR-FREE, OR THAT DEFECTS IN THE COURSEWARE WILL BE CORRECTED. FURTHERMORE, OSMA DOES NOT WARRANT OR MAKE ANY REPRESENTATIONS REGARDING THE USE OR THE RESULTS OF THE USE OF THE COURSEWARE IN TERMS OF CORRECTNESS, ACCURACY OR RELIABILITY, OR OTHERWISE. NO VERBAL OR WRITTEN INFORMATION OR ADVICE GIVEN BY OSMA OR ITS AUTHORIZED REPRESENTATIVE SHALL CREATE A WARRANTY OR IN ANY WAY INCREASE THE SCOPE OF THIS WARRANTY.

9. Limitation of Liability. TO THE MAXIMUM EXTENT PERMITTED UNDER APPLICABLE LAWS, UNDER NO CIRCUMSTANCES, INCLUDING NEGLIGENCE, SHALL OSMA, ITS AFFILIATES OR THEIR DIRECTORS, OFFICERS, EMPLOYEES OR AGENTS, BE LIABLE FOR ANY INCIDENTAL, SPECIAL OR CONSEQUENTIAL DAMAGES (INCLUDING DAMAGES FOR LOSS OF BUSINESS PROFITS, BUSINESS INTERRUPTION, LOSS OF BUSINESS INFORMATION AND THE LIKE) ARISING OUT OF THE USE OR INABILITY TO USE THE COURSEWARE, EVEN IF OSMA OR ITS AUTHORIZED REPRESENTATIVE HAS BEEN ADVISED OF THE POSSIBILITY OF SUCH DAMAGES. SOME JURISDICTIONS DO NOT ALLOW THE LIMITATION OR EXCLUSION OF LIABILITY FOR INCIDENTAL OR CONSEQUENTIAL DAMAGES SO THE ABOVE LIMITATION OR EXCLUSION MAY NOT APPLY. In no event shall OSMA's total liability to you for all damages, losses, and causes of action (whether in contract, tort, including negligence, or otherwise) exceed the greater of the amount paid by you for the courseware or U.S. $10.

10. Government End Users. If you are acquiring the courseware on behalf of any unit or agency of the United States Government, the following provisions apply. The enclosed courseware and related documentation are provided with "restricted rights". Use, duplication, or disclosure by the U.S. Government or any agency or instrumentality thereof is subject to restrictions as set forth in subdivision (c)(1)(ii) of the Rights in Technical Data and Computer courseware clause at 48 C.F.R. 252.227-7013, or in subdivision (c)(1) and (2) of the Commercial Computer Courseware-Restricted Rights Clause at 48 C.F.R. 52.22719, as applicable.

11. Controlling Law and Severability. This license shall be governed by and construed in accordance with the laws of the United States and the State of Delaware, as applied to agreements entered and to be performed entirely within Delaware between Delaware residents. If for any reason a court of competent jurisdiction finds any provision of this license, or portion thereof, to be unenforceable, that provision of the license shall be enforced to the maximum extent permissible to affect the intent of the parties, and the remainder of this license shall continue in full force and effect.

12. Entire Agreement. This license constitutes the entire agreement between the parties with respect to the use of the courseware and related documentation, and supersedes all prior or contemporaneous understandings or agreements, written or oral, regarding such subject matter. Any additional or different terms or conditions proposed by you or contained in any purchase order are hereby rejected and shall be of no force and effect unless expressly agreed to in writing by OSMA. No amendment to or modification of this license will be binding unless in writing and signed by a duly authorized representative of OSMA.

Legal Notices

Open Service Management (OSM®) Foundation

About this course

This 2-day (14 hour) course covers the key concepts, principles and models of IT Service Management and prepares learners to pass the OSM® Foundation Certificate in IT Service Management examination. The purpose of the OSM® Foundation Certificate in IT Service Management is to certify that the candidate has gained knowledge of the OSM® terminology, structure and basic concepts and has comprehended the core principles of OSM® practices for service management. The OSM® Foundation certificate in IT Service Management is not intended to enable holders of the certificate to apply the OSM® practices for service management without further guidance.

Certification Examination Type and Duration

Upon completion of the course, candidates may sit the optional OSM® Foundation examination leading to the OSM® Foundation Certificate in IT Service Management. Sixty (60) minutes (75 minutes and use of a dictionary for those taking the examination in a language other than their first language) is allowed for the closed-book, supervised examination which may be online or paper-based. The examination is a multiple-choice format consisting of forty (40) questions. A score of 26 out of 40 (65%) or better is required to pass the examination.

Audience

The intended audience of the OSM Foundation certificate in IT Service Management includes:

- Individuals who need a basic understanding of service management concepts strongly aligned to ISO 20000 and compatible with traditional IT service management guidance.
- Individuals responsible for service management who seek a comprehensive but agile and lightweight approach, who value open content they can contribute to and benefit from.
- Individuals in organizations who want to snap their internal frameworks to open, community driven guidance they can shape, rather than simply "rolling their own", or snapping to closed, proprietary standards.

Prerequisites

This course has no specific prerequisites.

Duration

2 days (14 hours).

Learning Objectives

After completing this course, learners will be able to list and describe the following:
- The concept of service management; structure and rationale for Open Service Management
- Stakeholder concepts, desired states and practices
- Value flow concepts, desired states and practices
- Services concepts, desired states and practices
- Service management system concepts, desired states and practices (for practices, list only)

Course Outline

Module 1: Service management and Open Service Management

Module 2: Stakeholder concepts, desired states and practices

Module 3: Value and value flow concepts, desired states and practices

Module 4: Service concepts, desired states and practices

Module 5: SMS concepts, desired states and practices

Module 6: Summary and exam preparation

Open Service Management®

OPEN COMMUNITY-DRIVEN BEST PRACTICE

Professional Qualifications for Open Service Management®

The Open Service Management Foundation Syllabus
Version 1.0, January 1, 2018

Certification Examination Format

This syllabus is for training for an accompanying certification examination, with the following format:

Type	Multiple choice, 40 questions.
Duration	Maximum 60 minutes for all candidates in their respective language, and an additional 15 minutes for candidates where the exam is not in their mother tongue.
Prerequisite	Certificate of completion from an Open Service Management Foundation Registered Training Organization (RTO) and Registered Trainer (RT).
Supervised	Yes
Format	Closed book.
Passing Score	26/40 or 65%
Delivery	This examination is available online or in paper format from Acquiros, the Official Examination Institute for Open Service Management.

THE OPEN SERVICE MANAGEMENT FOUNDATION CERTIFICATE

The purpose of the OSM Foundation certificate in IT Service Management is to certify that the candidate can describe Open Service Management essential concepts, including structure, terminology, definitions, desired states and practices, as a basis for applying these concepts.

Audience

The intended audience of the OSM Foundation certificate in IT Service Management includes:

- Individuals who need a basic understanding of service management concepts that are strongly aligned to ISO 20000 and compatible with traditional IT service management guidance.
- Individuals responsible for service management who seek a comprehensive but agile and lightweight approach, who value open content that they can contribute to and benefit from.
- Individuals in organizations who want to snap their internal frameworks to open, community driven guidance they can shape, rather than simply "rolling their own", or snapping to closed, proprietary standards.

Learning Objectives

At the end of their studies, Candidates should be able to list and describe the following:

- The concept of service management, and structure and rationale for Open Service Management
- Stakeholder concepts, desired states and practices
- Value flow concepts, desired states and practices
- Services concepts, desired states and practices
- Service management system concepts, desired states and practices (for practices, list only)

Note: OSM is open content, and the OSM Alliance is a non-profit advancing the compilation of open practices. If you are a subject matter expert and see something that isn't quite right, please consider contributing. We welcome all talents, resources and help. Sign up on osmalliance.org or contact us at info@osmalliance.org. Together, we can make practices for the people, by the people, in an agile way!

Contact Hours

The Open Service Management (OSM) Foundation certificate requires 14 contact hours of study, the equivalent of two traditional classroom training days with the examination at the end of the second day.

Foundation syllabus

This syllabus is intended to guide candidates in preparing for the OSM Foundation examination, and training organizations in the development and delivery of OSM Foundation training materials. The source content for this syllabus is the Open Service Management Body of Knowledge.

Candidates for the OSM Foundation certification examination must complete all units of the training and successfully pass the corresponding examination to achieve certification.

The table that follows lists the units and topics to be covered. Training organizations are expected to add value through activities and examples that drive the learning home. While training organizations may re-order the content as they see fit, they are expected to index all training content against the syllabus to assure trainees that all topics are covered, and to allow for ease of navigation of training content by syllabus reference and glossary term.

Unit	Content
OSMFND01	**Introduction to service management and Open Service Management (OSM)** The purpose of this unit is to help the candidate define core service management and OSM concepts, including things worth managing, desired states, best practices, stakeholders, value and value flow, services, service management, and service management system, and to explain the structure, rationale, and training and certification path for OSM. Specifically, candidates must be able to: 01-01. Explain what things worth managing are 01-02. Explain what desired states and force field analysis are 01-03. Explain what best practices are, the three types: deterministic, adaptive, emergent, and the two categories: mode 1 and mode 2 01-04. Explain how things worth managing, desired states, and practices relate 01-05. List and describe the stakeholders of service management 01-06. Describe what value and value flow are, and their characteristics 01-07. Define what a service is, and its characteristics 01-08. Define what service management is, and its characteristics 01-09. Define what a service management system is, and its characteristics 01-10. Explain what Open Service Management is, its characteristics and structure 01-11. Explain the training and certification path for OSM (non-examinable) **The recommended study period for this unit is minimum of 60 minutes, or 1 hour.**
OSMFND02	**Stakeholder concepts, desired states and practices** The purpose of this unit is to help the candidate list and describe the stakeholders of service management, their desired states, and best practices for achieving them, including the following stakeholders: 02-01. Stakeholders – Overall definition, desired state, best practices 02-02. Stakeholder customer – Definition, desired state, best practices 02-03. Stakeholder user – Definition, desired state, best practices 02-04. Stakeholder provider – Definition, desired state, best practices 02-05. Stakeholder supplier – Definition, desired state, best practices **The recommended study period for this unit is minimum 60 minutes or 1 hour.**
OSMFND03	**Value flow concepts, desired states and practices** The purpose of this unit is to help the candidate describe value flow through services, its desired state, and best practices for achieving it. 03-01. Define what value is, and its characteristics 03-02. Describe value flow in relation to value 03-03. Define value stream, and describe value stream mapping 03-04. Describe Lean, its relation to value, and the five lean principles 03-05. Describe Agile, its relation to value, and the four agile statements of value 03-06. Value flow – Overall definition, desired state, best practices **The recommended study period for this unit is minimum 60 minutes, or 1 hour.**

OSMFND04	Service concepts, desired states and practices

The purpose of this unit is to help the candidate list and describe the components that make up the configuration of services, their desired states, and best practices for achieving them, including:

04-01: Service concepts, desired states and practices - Configuration

04-01-01. Service – Overall definition, desired state, best practices
04-01-02. Service Configuration – Definition, desired state, best practices
04-01-03. Service Configuration – Human-Led Services – Definition, desired state, best practices
04-01-04. Service Configuration – IT-Led Services – Definition, desired state, best practices
04-01-05. Service Configuration – Software – Definition, desired state, best practices
04-01-06. Service Configuration – Applications – Definition, desired state, best practices
04-01-07. Service Configuration – Data – Definition, desired state, best practices
04-01-08. Service Configuration – Platform – Definition, desired state, best practices
04-01-09. Service Configuration – Runtime – Definition, desired state, best practices
04-01-10. Service Configuration – Middleware – Definition, desired state, best practices
04-01-11. Service Configuration – Operating System – Definition, desired state, best practices
04-01-12. Service Configuration – Infrastructure – Definition, desired state, best practices
04-01-13. Service Configuration – Virtualization – Definition, desired state, best practices
04-01-14. Service Configuration – Server – Definition, desired state, best practices
04-01-15. Service Configuration – Storage – Definition, desired state, best practices
04-01-16. Service Configuration – Network – Definition, desired state, best practices
04-01-17. Service Configuration – Hardware – Definition, desired state, best practices
04-01-18. Service Configuration – Facilities – Definition, desired state, best practices

04-02: Service concepts, desired states and practices - Functionality and Qualities

The purpose of this unit is to help the candidate list and describe the functional and non-functional features and characteristics of services, their desired states, and best practices for achieving them, including:

04-02-01. Service Functionality – Definition, desired state, best practices
04-02-02. Service Qualities – Definition, desired state, best practices

04-02-03 - 04-02-08: Service Qualities - Human-Led Services

04-02-03. Service Qualities – Human-Led Services - Definition, desired state, best practices
04-02-04. Service quality – Reliability – Definition, desired state, best practices
04-02-05. Service quality – Responsiveness – Definition, desired state, best practices
04-02-06. Service quality – Assurance – Definition, desired state, best practices
04-02-07. Service quality – Empathy – Definition, desired state, best practices
04-02-08. Service quality – Tangibles – Definition, desired state, best practices

04-02-09 - 04-02-32 Service Qualities - IT-Led Services

	04-02-09. Service Qualities – IT-Led Services – Definition, desired state, best practice 04-02-10. Service quality – Availability – Definition, desired state, best practices 04-02-11. Service quality – Manageability – Definition, desired state, best practices 04-02-12. Service quality – Serviceability – Definition, desired state, best practices 04-02-13. Service quality – Performance – Definition, desired state, best practices 04-02-14. Service quality – Reliability – Definition, desired state, best practices 04-02-15. Service quality – Recoverability – Definition, desired state, best practices 04-02-16. Service quality – Discoverability – Definition, desired state, best practices 04-02-17. Service quality – Assurance – Definition, desired state, best practices 04-02-18. Service quality – Security – Definition, desired state, best practices 04-02-19. Service quality – Integrity – Definition, desired state, best practices 04-02-20. Service quality – Credibility – Definition, desired state, best practices 04-02-21. Service quality – Compliance – Definition, desired state, best practices 04-02-22. Service quality – Usability – Definition, desired state, best practices 04-02-23. Service quality – Internationalization – Definition, desired state, best practices 04-02-24. Service quality – Accessibility – Definition, desired state, best practices 04-02-25. Service quality – Adaptability – Definition, desired state, best practices 04-02-26. Service quality – Interoperability – Definition, desired state, best practices 04-02-27. Service quality – Scalability – Definition, desired state, best practices 04-02-28. Service quality – Elasticity – Definition, desired state, best practices 04-02-29. Service quality – Portability – Definition, desired state, best practices 04-02-30. Service quality – Extensibility – Definition, desired state, best practices **The recommended study period for this unit is minimum 240 minutes, or 4 hours.**
OSMFND05	**Service management system (SMS) concepts, desired states and practices** The purpose of this unit is to help the candidate list and describe the what a service management system is, its typical components, their desired states, and best practices for achieving them, including the following SMS components: **05-01: Service management system (SMS) concepts, desired states and practices** 05-01-01. Service Management System (SMS) – Definition, desired state, best practices **05-02: SMS concepts, desired states and practices - Design & transition** 05-02-01. SMS – Design & transition – Overall definition, desired state, best practices 05-02-02. SMS – Design & transition- Strategy & GRC – Definition, desired state, best practices 05-02-03. SMS – Design & transition - Learning & improvement – Definition, desired state, best practices 05-02-04. SMS – Design & transition - Development – Definition, desired state, best practices 05-02-05. SMS – Design & transition - Release & deployment – Definition, desired state, best practices 05-02-06. SMS – Design & transition - Change & configuration – Definition, desired state, best practices **05-03: SMS concepts, desired states and practices - Promotion**

	05-03-01. SMS – Promotion – Overall definition, desired state, best practices **05-04: SMS concepts, desired states and practices - Support** 05-04-01. SMS – Support – Overall definition, desired state, best practices 05-04-02. SMS – Support – Event handling – Definition, desired state, best practices 05-04-03. SMS – Support – Incident handling – Definition, desired state, best practices 05-04-04. SMS – Support – Major incident handling – Definition, desired state, best practices 05-04-05. SMS – Support – Problem handling – Definition, desired state, best practices 05-04-06. SMS – Support – Request handling – Definition, desired state, best practices **05-05: SMS concepts, desired states and practices - Delivery** 05-05-01. SMS – Delivery – Overall definition, desired state, best practices 05-05-02. SMS – Delivery – Stakeholder relations – Definition, desired state, best practices 05-05-03. SMS – Delivery – Administration – Definition, desired state, best practices 05-05-04. SMS – Delivery – Provisioning, metering, billing – Definition, desired state, best practices 05-05-05. SMS – Delivery – Budgeting & accounting – Definition, desired state, best practices **The recommended study period for this unit is minimum 300 minutes, or 5 hours.**
OSMFND06	**Sample examination revision** The purpose of this unit is to help the candidate to pass the OSM Foundation examination. Specifically, candidates must: 06-01. Sit a minimum of one OSM Foundation mock examination. **The recommended study period for this unit is minimum 120 minutes, or 2 hours including examination revision.**

2018 edition

Course 57011A | v1.0.0

OSM® Foundation

Based on the OSM® Foundation Certificate Syllabus v1.0

The OSM® Alliance
Open community-driven best practices

Open Service Management® Foundation is a 2-day (14 contact hour) course that teaches the essential concepts of Open Service Management, or OSM® (pronounced, "awesome"), which is brought to you by the Open Service Management Alliance, or OSMA, an non-profit based in Seattle, Washington USA . The OSMA's mission is to be an open and agile platform for fresh, relevant, lightweight community-drive best practice guidance, where best practice is defined as, "what people are doing now that works".

This OSM Foundation course is based on the first iteration of OSM best practices. The structure and terminology of OSM best practices are strongly aligned with ISO 20000, so as to be compatible with traditional ITSM guidance, but varies from it as it is written for a target environment of a hybrid of traditional IT and cloud, and its associated practices.

This course is meant to prepare you for the OSM Foundation examination, which is a 1-hour, 40-question, multiple-choice exam with 26 of 40 (65%) of questions required to pass. To register for the exam, you must have taken a course (which can include self-study), from a Registered Trainer and Registered Training Organization.

This course is intended for individuals, teams and organizations that are looking to snap to a lightweight, open framework that is free for end users to use, and community-driven, and developed in an agile style, with the idea that an open, agile, community-driven approach is something worth benefiting from and contributing to, as it will drive fresh, relevant content and ubiquity of use; it is for those who'd prefer this over "rolling their own", at great time and cost, with 100% assurance that what they roll will not map to that of the next individual, team, or organization; and it is for those who do not wish to snap to heavyweight, proprietary guidance that is closed to community contribution.

To sign up to contribute to OSM, either with your time and expertise or with funding, please visit osmalliance.org or write the OSMA at info@osmalliance.org.

OSM Foundation Course Agenda

Module 1: Service management and Open Service Management	60m
Module 2: Stakeholder concepts, desired states and practices	60m
Module 3: Value and value flow concepts, desired states and practices	60m
Module 4: Service concepts, desired states and practices	240m
Module 5: SMS concepts, desired states and practices	300m
Module 6: Summary and exam preparation	120m

The minimum contact hours for this course are 14, including 2 hours for sample examination revision.

This course is divided into six main topics or modules:

1. Module 1 introduces the concept of services and service management in general, and the structure and core concepts of Open Service Management in particular. (60m)

OSM has four components parts to its structure, four, "things worth managing", and the next four modules cover those parts. In each, core concepts are covered, along with desired states (basically, what good looks like), and best practices.

2. Module 2 covers stakeholders. (60m)

3. Module 3 covers value and value flow. (60m)

4. Module 4 covers services, including their parts (or configuration), features (or functionality), and qualities (or non-functional characteristics). (240m)

5. Module 5 covers the service management system (or SMS), which is your mechanism to ensure value flows through services to stakeholders. (300m)

6. Module 6 summarizes the course and prepares you to take the OSM Foundation certification examination. (120m)

OSM Foundation Course Objectives

At the end of the course, you should be able to describe the core principles, practices, and terminology associated with service management, including:

- Service management and Open Service Management
- Stakeholder concepts, desired states and practices
- Value and value flow concepts, desired states and practices
- Service concepts, desired states and practices, including service configuration, functionality and qualities
- Service Management System (SMS) concepts, desired states and practices, including Design & transition, promotion, support and delivery

This is what the Foundation examination will cover. Non-examinable topics in this course will be marked with the icon:

Module 1
Service management and Open Service Management (60m)

This is the first module of the course. Here we introduce the concept of services and service management, and the core concepts and structure of Open Service Management.

OSM Foundation Course Agenda

Module 1: Service management and Open Service Management	60m
Module 2: Stakeholder concepts, desired states and practices	60m
Module 3: Value and value flow concepts, desired states and practices	60m
Module 4: Service concepts, desired states and practices	240m
Module 5: SMS concepts, desired states and practices	300m
Module 6: Summary and exam preparation	120m

The minimum contact hours for this course are 14, including 2 hours for sample examination revision.

In this module, we'll cover concept of services and service management in general, and the structure and core concepts of Open Service Management in particular. So let's get started!

Module Objectives

At the end of this module, you should be able to to define core service management and OSM concepts, including things worth managing, desired states, best practices, stakeholders, value and value flow, services, service management, and the service management system, and explain the training and certification path for OSM. Specifically, you should be able to:

01-01. Explain what things worth managing are

01-02. Explain what desired states and force field analysis are

01-03. Explain what best practices are, and the three types: deterministic, adaptive, emergent, and two categories: mode 1 and mode 2

01-04. Explain how things worth managing, desired states, and practices relate

01-05. List and describe the stakeholders of service management

01-06. Define what value and value flow are, and their characteristics

01-07. Define what a service is, and its characteristics

01-08. Define what service management is, and its characteristics

01-09. Define what a service management system is, and its characteristics

01-10. Explain what Open Service Management is, its characteristics and structure

01-11. Explain the training and certification path for OSM (non-examinable)(non-examinable)

The recommended study period for this unit is minimum of 60 minutes, or 1 hour.

What specific objectives do you have for these topics?

What are "Things worth managing?"

1. Stakeholders	Customers, users, the provider, and suppliers
2. Value and value flow	To stakeholders, multidirectionally, through services
3. Services	Including configuration (parts), functionality (features) and qualities (behaviors)
4. Service Management System	The mechanism the provider uses to direct and control services for sustainable, continuous value flow to stakeholders

Figure 01-01.1 Things worth managing
01-01. Explain what things worth managing are

For all things worth doing, there are "things worth managing" for which we must aim to achieve and sustainably maintain a desired state. Service management is no exception; it has four, "things worth managing": Stakeholders, value and its flow, services, and the service management system or SMS.

OSM seeks to compile the generic set of things worth managing common to all providers as a starting point for desired state-based management of services. For every service provider, these are the generic set of things worth managing.

In addition, there are typically additional things worth managing due to the provider's specific situation, tools used, and so on. The idea is that each service provider needs to identify their "things worth managing" and work to keep them in a "green" or "good" state.

What is a desired state?

Desired state

The state of "goodness" to achieve and sustainably maintain, the continuous "end in mind" for things worth managing; not an output or deliverable or outcome, although these could contribute to (or detract from) a desired state.

- There are two main approaches to managing things worth managing: focusing on desired states ("ends"), or "means"; e.g., seeking to "minimize the business disruption of change, and know what changed" is a desired state or "end"; you might focus on this, or on "means"—e.g., engineering processes, etc.; OSM describes means (best practices), but focuses on ends
- Many things can be barriers or enablers to achieving and maintaining desired states, e.g., a process or tool can be a means to achieving and maintaining desired states, or it can be a blocker, or some combination thereof, depending on how it was designed, tools used, organizational culture, cadence, etc.
- In OSM, desired states are written from the perspective of the service provider

01-02. Explain what desired states and force field analysis are

At its inception, with Jan Carlzon's "Moments of Truth", service management was lightweight, agile, and focused on outcomes, or ends, and interactions among people ("moments of truth").

And while traditional ITSM guidance seeks be outcome based, by focusing on, "What to do and why", and giving an EXAMPLE of how, quite often in practice people have applied this guidance with a focus on a heavy-handed how, layering heavyweight, end to end processes on top of the work they do, without a call to action for individuals, teams, and the organization to know, and drive towards, service management outcomes.

OSM returns to the roots of service management, and puts its focus squarely on outcomes, on maintain desired states for things worth managing. So for example, "changes" are a "thing worth managing", with an outcome of, "we minimize the business disruption of change, and can track what changed"; a process could enable—or disable—this desired state.

OSM focuses squarely on these outcomes, with a call to action for individuals and teams in organizations to know the things worth managing, and their desired states, and to work within their patch to drive to achieve and maintain desired states for things worth managing.

The idea is to make knowing and driving towards desired states, by working to remove and reduce barriers and add and amplify enablers to achieving and maintaining those desired states, "how we do things around here".

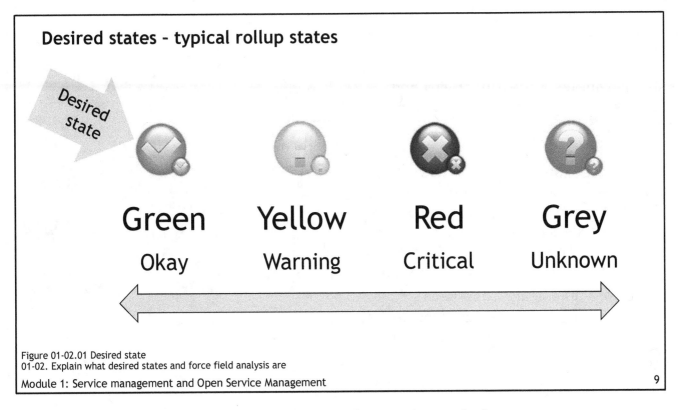

Figure 01-02.01 Desired state
01-02. Explain what desired states and force field analysis are
Module 1: Service management and Open Service Management

9

Let's talk about the different states something worth managing can be in.

There are typically four states tracked in modern monitoring systems:

1. Green means good, it's in the desired state.

2. Yellow means potentially or actually degraded in some way, for example, disk space is filling up or circuit utilization is high, which may be affecting performance.

3. Red means bad, dead, down, or critical.

4. Grey means "unknown"—the thing worth managing is in an unknown state.

Think of it this way, if you join or start a service provider organization, and are responsible for managing it, you'll want to know a couple of things.

First, what is "it"? What are the things worth managing that you are responsible for?

Once you know what "it" is, your next concern will be anything that's red, or critical—how do we get that back online, to green?

Next, you'll want to address anything that's degraded or potentially degraded—the yellow stuff.

And lastly, anything that's in an unknow state, you'll want to get to a known state, and then work towards making whatever is red or yellow, green.

Force field analysis

Force field analysis, developed by Kurt Lewin, helps identify driving and restraining forces you must amplify and reduce to achieve a desired state

Here's how: brainstorm enablers (driving forces), and barriers (restraining forces); identify actions to add or amplify enablers / remove or minimize barriers; prioritize based on effort and impact; pick a subset to go after to move towards the desired state

Force field analysis is a key technique for achieving and maintaining desires states for things worth managing

Driving Forces (enablers for change)
+ and ↑

Restraining Forces (obstacles to change)
- and ↓

Desired State

Source: http://www.change-management-coach.com/force-field-analysis.html

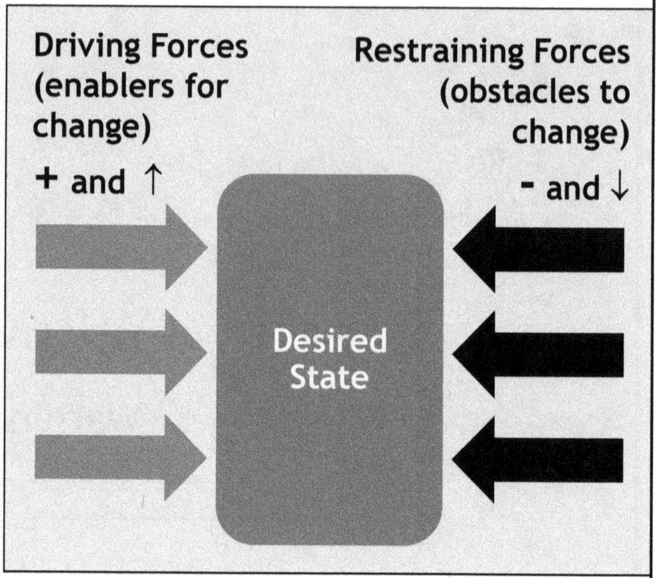

Figure 01-02.2 Force field analysis
01-02. Explain what desired states and force field analysis are

So let's say you've identified your things worth managing, and a desired state for each that you want to achieve and maintain, sustainably. What's a simple technique you can use to do that?

The most universally applicable technique is Kurt Lewin's force field analysis. It works like this:

- You are "here". You'd like to be somewhere else. To get there, you ask, "what forces can we add or amplify to drive towards our goal?"; "what barriers are between us and our goal, and how can we minimize or eliminate them?"

- You then look at the barriers and enablers and take action to minimize or eliminate barriers, and add or amplify enablers, putting your resources to work on the subset of these that will give you the greatest "bang for your buck".

This is precisely what we are doing when, for example, we put build pipelines and continuous integration and delivery and the like in place, and work to continuously improve the value flow we get from them. We are removing barriers and amplifying enablers to increase flow.

And in today's organizations, we all are responsible for doing so—we all own the desired states, and we all have a an active role in achieving and maintaining them.

What are best practices?

- Best practices are things people are actually commonly doing now that work well when aiming to achieve and maintain desired states / desired states for things worthy of managing
- Best practices are not bleeding, or leading-edge practices; they are practices people are currently using that work across multiple organization, and are producing successful results

- Like traditional ITSM guidance, OSM does not create best practices, but seeks to provide a platform that gives the IT community a reason to compile and maintain a source of current best practices, within a common structure that makes finding, understanding, applying, and improving them easier

01-03. Explain what best practices are, and the three types: deterministic, adaptive, and emergent and two categories: mode 1 and mode 2

Module 1: Service management and Open Service Management

11

Traditional ITSM guidance talk about best practices as being commodity practices, not bleeding, or leading edge, which is true; best practices should be common practices.

The trouble with some traditional ITSM guidance is that because it is not updated regularly, or managed in a data-driven way, it can devolve into, "what people said they used to do that worked ten years ago when this was written, using tools and a target environment that no longer exists".

OSM defines best practices as, "what people are doing now that works". This is important. Like more traditional ITSM frameworks, OSM compiles, rather than creates, best practice.

Unlike more traditional ITSM frameworks, OSM endeavors to do this in an agile way, driven by community contributions, to ensure fresh, relevant content, updated more frequently, and curated using facts to ensure what is cited as best practices are in fact in wide use.

Now lets' talk about the three types and two modes of best practice that OSM defines.

- There are three broad categories of practices—deterministic (like McDonalds), adaptive (like Agile / Scrum), and emergent (completely new-to-you, e.g., the first time doing container orchestration—there is no precedent)
- All three are valid and can be enablers in some circumstance and barriers in others, or some combination of both; your goal should be to make sure there is a good match between the type of practices and your situation, and to be open to changing based on changing circumstances, e.g., target environment

Deterministic	You will do these things—it's not negotiable/this is designed in to our systems, tools, artifacts in advance, e.g., bug-tracking system
Adaptive	We have milestones / buckets of activities, but sequence, which ones, by whom, etc. isn't determinable in advance—we adapt to the situation
Emergent	Nothing predefined—practices emerge in the situation

Figure 01-03.1 Three broad categories of practices

01-03. Explain what best practices are, and the three types: deterministic, adaptive, and emergent and two categories: mode 1 and mode 2

OSM adopts the three types of practices David Pultorak (source: Rethinking BPM: Business Processes as Networks of Human Interactions, Business Process Management Conference 2005 Thursday, June 16, 2005, The Palmer House Hilton, Chicago, IL) defines: deterministic, adaptive, and emergent. This may sound like a fancy, academic categorization scheme, but the basic idea is simple:

- First, if you're going to do something over and over again, and want consistent results, you'll want to make it like a machine, or a coocoo clock—wind it up and it always produces the same result—like McDonald's, those chicken nuggets always come out in so many seconds, and always look and taste like this. That's deterministic. It's appropriate for such situations.

- Another situation is where what you actually do—the activities you do, in what sequence, with what tools, for what duration, etc.—should change based on feedback loops. This is the domain of Agile Scrum projects. In this scenario, you may have modules or routines or patterns you pull from, but precisely who will do what, when, will be driven by the situation.

- Lastly, you have emergent practices, like when container orchestration is a new thing and it's the very first time you and others are trying it. There is no precedent, no roadmap, you are literally making it up as you go, with patterns and practices emerging along the way.

Why make these distinctions? Because quite often, in heavyweight implementations of service management, most or all emphasis is put on setting up deterministic, end-to-end processes, with no room left for what is adaptive or emergent. This is a problem because the deterministic scenario reflects only a part of the reality we are faced with. OSM attempts to correct this by "shifting left" towards innovation and allowing room for all three modes of practice, as all three can be valid, and the right one to choose depends on the situation at hand.

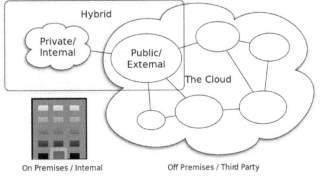
In addition to three types of practices—deterministic, adaptive, and emergent, there are two modes of practices, as driven by the target environment you manage.

Gartner calls these "mode 1" and "mode 2", and this language seems as good as any, so we'll use it here.

- "Mode 1" practices are ideally suited to traditional IT environments, things like traditional IT configuration management, where we work hard to have a good logical picture of the components of our environment, along with how they relate to one another.

- "Mode 2" practices are ideally suited to cloud / mobile environments, things like infrastructure-as-code and immutable deployments, where, for example, the emphasis shifts in configuration management to making sure the code that is used to create things is good, because we know what things look like, as they are uniform (as they are "cattle", not "pets", as DevOps practitioners like to say).

Today, the typical IT environment is a hybrid of cloud and mobile; therefore the typical practices are (or should be) an appropriate blend of mode 1, 2, and shared practices.

Gartner go on to contrast mode 1 and mode practices based on their characterizing metaphor, operating model, overall goal, what is valued, the approach taken, governance model, sourcing model, talent needed, cultural characteristics, and typical work cycle.

OSM attempts to catalog mode 1, 2, and shared best practices. The practices cited are meant to be examples, not definitive lists, complete and valid for all time.

At this, the foundation level, all that is required is that you be able to recall a few best practices associated with each "thing worth managing". The idea is that you can draw a line between things worth managing, their desired states, and some ways (best practices) to achieve and maintain them. It is not necessary at the foundation level to be able to define each best practice, or to be able to put it into practice—this is for the practitioner level of study and practice.

How things worth managing, desired states, and best practices relate

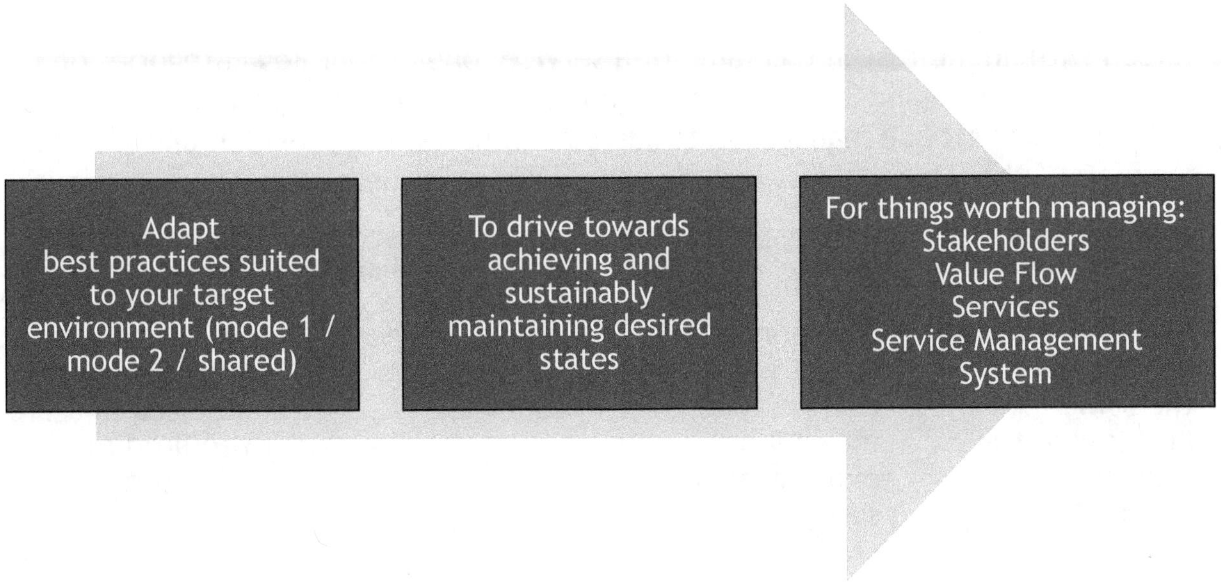

| Adapt best practices suited to your target environment (mode 1 / mode 2 / shared) | To drive towards achieving and sustainably maintaining desired states | For things worth managing: Stakeholders Value Flow Services Service Management System |

Figure 01-04.1 How things worth managing, desired states and practices relate
01-04. Explain how things worth managing, desired states, and practices relate

So OSM seeks to compile current best practices, both mode 1 and 2, as well as shared practices applicable to both traditional IT and cloud environments.

The idea is that you can apply the subset and blend of these that are suited to your environment, as a basis for driving towards and achieving and sustainably maintaining desired states, for things worth managing.

The four stakeholders of service management

Stakeholder

Person with an interest in the value of a service; one of four key categories of things worth managing in service management, there are four primary stakeholder in service management: customers, users, the provider, and suppliers. Other stakeholders include shareholders, owners, etc.

Customers	Users	Providers	Suppliers
Those who pay for services; may be internal (in the same business as the provider) or external	Those who use the service, but do not pay for it; users may be internal or external to a service provider	Entities that deliver services to customers and users	Third parties who supply providers with services that are required to deliver provider services

Figure 01-05.1 Four stakeholders in service management
01-05. List and describe the stakeholders of service management

Module 1: Service management and Open Service Management

16

Stakeholders are one of the four key "things worth managing" defined by OSM .

Here you should see some similarities and a key difference here between traditional ITSM guidance and OSM.

First, the similarities: traditional ITSM guidance cites customers, users, and suppliers as stakeholders.

Now for the key difference: traditional ITSM guidance does not include the provider as a stakeholder in service management. Perhaps because it is assumed that the provider is THE key stakeholder.

In OSM, we list the provider as a stakeholder, as the provider certainly does have a stake in the key outcomes of service management.

What is value? And where does it reside?

Value (of a service)

The worth, in the mind of stakeholders, of a product or service, which is a function of the extent the stakeholder appraises it to fulfill a need or desire. Key components of value of a service include the price (for a customer) or profit (for a provider or supplier), functionality (features) and qualities of the service. Other components of value include enhanced market share, competitive advantage, capability, compliance, safety, reputation, goodwill, and learning. Value must flow multidirectionally, to and from all stakeholders, for services to be sustainable.

Since value is in the mind of stakeholders, as providers we must influence both the reality of our services and their perception by stakeholders

Source: Adapted from http://www.ame.org/sites/default/files/query_archive_docs/LeanGlossary_01_08_1.pdf
01-06. Define what value and value flow are, and their characteristics

Module 1: Service management and Open Service Management

Value and its flow is another "thing worth managing" in OSM.

Here again you should see some similarities and differences between OSM's conception of value, and that of tradition ITSM guidance.

First, the similarity: both traditional ITSM guidance are focused on value.

Now for the differences:

- Traditional ITSM guidance depicts value flowing unidirectionally from the provider to the pay-the-bills customer through services. In OSM, while the pay-the-bills customer is the PRIMARY arbiter of value (as without them, you cannot exist), all stakeholders—including the providers, suppliers, and users—have a stake in value, and a perspective that must be honored for sustainable success.

- In OSM, value is delivered not just through services and their features and qualities, and can take many forms. For example, value can flow back to the provider from customers and users in the form of feedback on services, feature sets, features, and new feature requests, and not just in the form of compensation for services rendered.

What is value flow?

Value Flow (of a service)

The multidirectional stream of value delivered through services to all stakeholders, primarily customers, users, the provider, and suppliers; value is created and added through design, transition, promotion, support, and delivery activities (which can be fully or partly automated or manual), as a applied to a service and its associates configuration, features, and qualities.

- Value flow is good when it proceeds to stakeholders with no undue stoppages, scrap, rework or backflows
- Barriers to flow include inconsistency and variation (Mura) and overburden or unreasonableness (Muri)
- Enablers to flow include just-in-time (JIT) systems and corrections for overburdened and unreasonable practices

01-06. Define what value and value flow are, and their characteristics

18

The flow of value is another thing worth managing in service management.

Here again, there are key differences between traditional ITSM guidance and OSM:

- OSM emphasized not just value, but on value flow, in line with agile and lean practices.

- OSM depicts value flow as multi-directional, among all stakeholders—customers, users, the provider, and suppliers—contrast this with traditional ITSM guidance, which depicts value as flowing in one direction, from the provider to the customer). In this view, beside providing value to customers, services must also provide value to users, and value back to the provider and its suppliers, in order for services to be sustainable.

All of these differences are important, especially in the cloud portion of today' typical hybrid environment. Teams are aligned around build pipelines and are working to make value flow unencumbers by inconsistency, variation and overburden. So they understand the need for value flow, and their role in making it happen, and improving it, and are creating and looking for best practices to help them do it.

What is a service?

Service

A means for a provider to deliver value to customers and users and return value back to the provider and its suppliers by facilitating desired states customers and users want to achieve and sustainably maintain; customers engage providers for services typically when doing so presents lower cost, effort and risk as compared to doing it themselves or alternatives; by doing so, customers aim to focus on ends / desired states versus details of the means, which the provider handles.

Services are one of four key categories of things worth managing in service management

Services can be internal or external, i.e., provided to customers and users in the same business entity as the provider; or to those outside of the provider's business entity

Customer	Service Provider
▪ Pays for and does not perform services	▪ Gets paid to perform services
▪ Owns costs and risks, but not <u>details</u>	▪ Takes on the <u>details</u> of costs and risks
▪ Accountable for ends / desired states	▪ Responsible for means to achieve desired states

Figure 01-07.1 Customer and service provider
01-07. Define what a service is, and its characteristics

Besides stakeholders and value and its flow, another thing worth managing in service management is services.

A service is a means of delivering value to customers by facilitating desired states customers want to achieve, at (what should be) a lower cost and risk than them doing it themselves, or through alternatives, and without the need to pay attention to the details (or the means) of delivering the service.

The general idea—and you are familiar with this for services you personally choose to use instead of simply doing things yourself, or using an alternative—is that you use a service because doing so lets you focus on the ends, or outcomes, instead of the means, and frees you from having to deal with the details associated with costs and risks, and because the service presents a better value in the end than doing it yourself or alternatives.

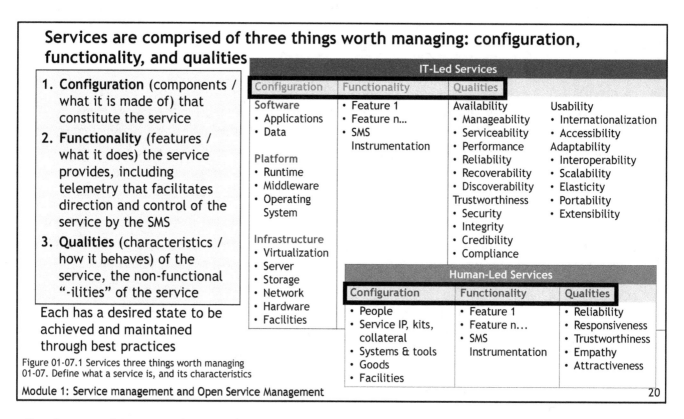

Services are comprised of three things worth managing: configuration, functionality, and qualities

1. **Configuration** (components / what it is made of) that constitute the service

2. **Functionality** (features / what it does) the service provides, including telemetry that facilitates direction and control of the service by the SMS

3. **Qualities** (characteristics / how it behaves) of the service, the non-functional "-ilities" of the service

Each has a desired state to be achieved and maintained through best practices

IT-Led Services			
Configuration	**Functionality**	**Qualities**	
Software • Applications • Data Platform • Runtime • Middleware • Operating System Infrastructure • Virtualization • Server • Storage • Network • Hardware • Facilities	• Feature 1 • Feature n... • SMS Instrumentation	Availability • Manageability • Serviceability • Performance • Reliability • Recoverability • Discoverability Trustworthiness • Security • Integrity • Credibility • Compliance	Usability • Internationalization • Accessibility Adaptability • Interoperability • Scalability • Elasticity • Portability • Extensibility

Human-Led Services		
Configuration	**Functionality**	**Qualities**
• People • Service IP, kits, collateral • Systems & tools • Goods • Facilities	• Feature 1 • Feature n... • SMS Instrumentation	• Reliability • Responsiveness • Trustworthiness • Empathy • Attractiveness

Figure 01-07.1 Services three things worth managing
01-07. Define what a service is, and its characteristics

Services are things worth managing that in turn are made up of three things worth managing:

1. Its components (that is, the service's parts, or what it is made up of; e.g., for IT-led services, infrastructure, platforms, software and the like)

2. The service's functionality (that is, the service's features, or what it does) that the service provides, including telemetry for direction and control by a Service Management System (or SMS), and the service's

3. Qualities (that is, its performance characteristics, or how it behaves), also known by developers as its non-functional "-ilities".

You can see here that OSM also defines two types of services (IT-led and Human-Led)—we'll get to the differences between these in a moment.

In the meantime, you should see that each of these have configuration, functionality, and qualities that must be managed.

This breakdown—configuration, functionality, and qualities—is another departure from traditional ITSM guidance. In traditional guidance, components of a service are depicted, but vary from what is depicted by OSM, which snaps to the infrastructure / platform / software stacks associated with cloud "as-a-service" architecture. Also, in traditional guidance, features are talked about as "utility" items, and a subset of operational qualities—availability, capacity, security, continuity, and supplier management—are discussed as "warranty" items. OSM expands the list of qualities to a "shift-left" list of qualities developers are concerned with.

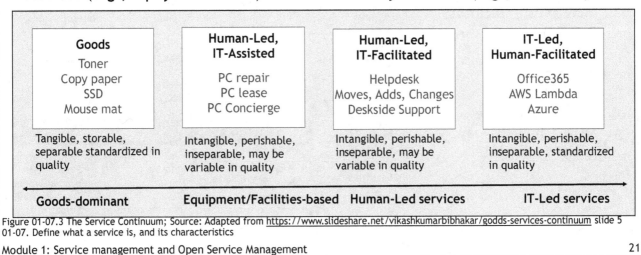

The service continuum – from goods to IT-Led services

A Service is some combination of an idea, physical good, or service a provider provides to customers and users; we are concerned with two types of services, Human-Led and IT-Led; we concerned with goods here only as components of services (e.g., a physical router) or as delivered by a service (e.g., PC install)

Goods	Human-Led, IT-Assisted	Human-Led, IT-Facilitated	IT-Led, Human-Facilitated
Toner Copy paper SSD Mouse mat	PC repair PC lease PC Concierge	Helpdesk Moves, Adds, Changes Deskside Support	Office365 AWS Lambda Azure
Tangible, storable, separable standardized in quality	Intangible, perishable, inseparable, may be variable in quality	Intangible, perishable, inseparable, may be variable in quality	Intangible, perishable, inseparable, standardized in quality

← Goods-dominant   Equipment/Facilities-based Human-Led services   IT-Led services →

Figure 01-07.3 The Service Continuum; Source: Adapted from https://www.slideshare.net/vikashkumarbibhakar/godds-services-continuum slide 5
01-07. Define what a service is, and its characteristics

OSM recognizes that we deliver value through products and services, and that there is a continuum. This acknowledgement—that we deliver value through products and services, and combinations of the two—is another key departure from traditional ITSM guidance.

A further departure is that OSM acknowledges two key types of services—again, along a continuum—namely, Human, and IT led. Human-Led services are led by humans, assisted in varying degrees by automation; IT-Led services (like Office365) are primarily represented by the automation, with humans assisting on the back end to varying degrees. The lack of this distinction in more traditional ITSM guidance led to someconfusion among practitioners around, for example, the nature of a service, and what to put in their service catalogs.

OSM recognizes that goods are different in nature from services. For example, a candy bar is different from a haircut, and a Bluetooth speaker is different from Office365. Generally, goods are more tangible, storage, separable, and come in some sort of standard format.

On the other end of the spectrum, IT services tend to delivered and consumed in the same moment; they are intangible, perishable, inseparable, and, like goods, standardized in quality.

And that's the point: there's a spectrum. We all have seen over the years how products have become service-like, and services have taken on product characteristics. It might be better to just call them all serv-octs or prodices, but all kidding aside, that's what we've got on our hands. For example, that piece of software running on your desktop—it's kind of a product, but now you download it digitally, you get updates, you get new features delivered in streams, you pay for it as a subscription—it's getting servicy.

This distinction, between human and IT-led services, is important because both Human-Led and IT-Led services are different in nature and have different characteristics to tend to if we are going to achieve and maintain a desired state for them.

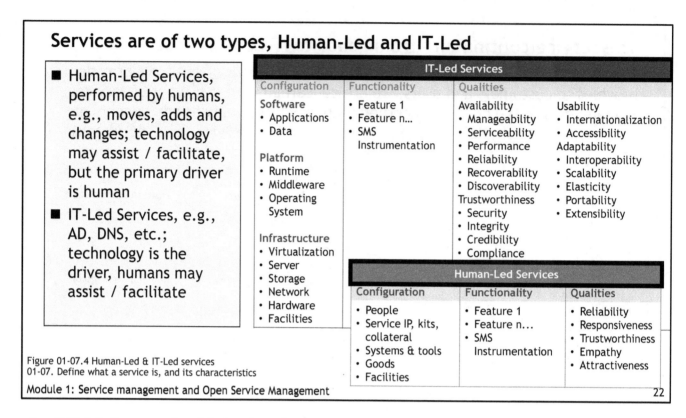

Services are of two types, Human-Led and IT-Led

- Human-Led Services, performed by humans, e.g., moves, adds and changes; technology may assist / facilitate, but the primary driver is human
- IT-Led Services, e.g., AD, DNS, etc.; technology is the driver, humans may assist / facilitate

IT-Led Services

Configuration	Functionality	Qualities	
Software • Applications • Data Platform • Runtime • Middleware • Operating System Infrastructure • Virtualization • Server • Storage • Network • Hardware • Facilities	• Feature 1 • Feature n... • SMS Instrumentation	Availability • Manageability • Serviceability • Performance • Reliability • Recoverability • Discoverability Trustworthiness • Security • Integrity • Credibility • Compliance	Usability • Internationalization • Accessibility Adaptability • Interoperability • Scalability • Elasticity • Portability • Extensibility

Human-Led Services

Configuration	Functionality	Qualities
• People • Service IP, kits, collateral • Systems & tools • Goods • Facilities	• Feature 1 • Feature n... • SMS Instrumentation	• Reliability • Responsiveness • Trustworthiness • Empathy • Attractiveness

Figure 01-07.4 Human-Led & IT-Led services
01-07. Define what a service is, and its characteristics

Module 1: Service management and Open Service Management   22

So, OSM defines two key kinds of services

1.   Human-Led services, performed by humans, e.g., moves, adds and changes; technology may be involved, but the primary driver is human, and

2.   IT-Led services, e.g., AD, DNS, etc.; technology is the driver, humans may be involved

Other characteristics of services

Can be provided in whole or in part by suppliers	Can be fully or partially automated or manual	Can involve provision of goods
Can be made up of sub-services, which can be shared	Can be broken down into types, e.g., core, optional, supporting	Can be stratified into tiers, e.g., gold, silver, bronze

Other characteristics of services include:

- Services can be provided in whole or in part by suppliers

- Service can be fully or partially automated or manual

- Services can involve the provision of goods, e.g., a replacement keyboard

- End-services are what that the customer recognizes and pay for; they typically include sub-services (e.g., updates, DHCP), which may be shared with other services.

- Services can be broken down into types, e.g.,

    o   Core services (e.g., for mobile phone service, dial tone),

    o   Optional services (e.g., data plan), and

    o   Supporting services (e.g. backups, update), and

- Services can be stratified in tiers (e.g., gold, silver, bronze level packages) that may feature different service configuration, functionality, and qualities.

So far we have covered stakeholders, value and its flow, and services as "things worth managing". The fourth and last "thing worth managing" is your system for managing services.

In service management, we (the provider), present ourselves to customers and users as a set of services. We align all things—our capabilities, resources, activities, and systems—up front, and over time and through changing circumstances—to ensure we can provide value consistently and sustainably, by having the wherewithal we need to consistently delivery on service commitments.

We do this by making explicit, discrete commitments; by defining what we will do, but also, what we won't do. The general posture towards customers and users should not be, "no", as this posture ensures customers and users don't get the value they need. It should also not be, "yes"— to everything, because this leads to everything being delivered by "best effort"—i.e., you literally cannot count on the provider. The right posture is, "yes we can do it, and this is what it will cost". In other words, a posture of listening to what is desired, and insisting on having the wherewithal to consistently deliver on it, or adjust commitments, to align to the wherewithal we can have.

In service management, we seek to align all our capabilities around services, including:

1. Technical systems – equipment, databases, software systems etc.

2. Managerial systems – for management of operations, including that of technical systems

3. Learning systems – for the maintenance of personal and team skills and knowledge

4. Cultural systems–for the regulation of values and norms, i.e., behaviors and objectives

The "bet" is that organizing around and presenting ourselves as a set of services is a superior operating model, that, given the same set of resources, we will do better than if we simply organized around the caring and feeding of technology.

What is a service management system (SMS)?

Service management system (SMS)

The dynamic mechanism, which may be fully or partially automated or manual, used in service management by a service provider to direct and control resources, activities and services to achieve and sustainably maintain the desired state of providing value to stakeholders in the form of services, over time and through changes to stakeholder needs and what technology makes possible. The SMS includes mechanisms to support how services are planned, designed, developed, deployed, promoted, supported, delivered, monitored, measured, reviewed, maintained, and improved.

Figure 01-09.1 Service Management System (SMS)
01-09. Define what a service management system is, and its characteristics

Module 1: Service management and Open Service Management

Service Management System (SMS)

Design & transition
- Alignment, Learning & improvement
- Development
- Release & deployment, change, retirement

Promotion

Support
- Event handling
- Request handling
- Incident handling
- Problem handling
- Major incident & disaster handling

Delivery
- Stakeholder relations
- Administration
- Provisioning, metering, billing
- Budgeting & accounting

25

The Service Management System, or SMS, may sound fancy or complicated, but it's really just the sum of whatever mechanisms you have in place to make sure stakeholders, value and its flow, and services are in their desired state, initially, and over time and changing circumstances.

Accordingly, the SMS must be a dynamic mechanism, because with the passing of time, stakeholder needs change, and new technologies create new possibilities, and so the value equation changes. As a result we must continuously improve services, by adding or changing services, feature sets and features, and retiring services that no longer add value.

As you can see in the figure, a service management system can be broken down into Design & transition, promotion, support, and delivery. OSM handles these a bit differently than traditional ITSM frameworks, which position these as sequential phases and processes.

In OSM, all of these are seen as going on simultaneously, not sequentially. In other words, while of course there is sequence in workflow, it is not helpful to, for example, put incident handling in one "phase" because "that's where it first becomes important".

OSM position these elements not as "processes", but as "things worth managing"—in other words, each represents a key outcome or desired state that we want to achieve and maintain. The focus here is on ends, not means, as we know that a process is just one means to achieve an outcome, and that a broken process (or a focus on process without attention to other enablers), can disable as well as enable sustained achievement of desired outcomes.

Further, in OSM, we seek to focus on outcomes, as opposed to activities, as this is more lightweight, and suited to today's environment and agile practices.

What is Open Service Management?

Open Service Management

An open, community-driven set of lightweight and agile best practice guidance to help practitioners create and enhance continuous, multi-direction value flow to stakeholders through services; value flow for stakeholders is a function of providing services with the right configuration, functionality and qualities, directed and controlled by a service management system (SMS) to achieve desired states for the things worth managing (stakeholders, value and value flow, services, and the SMS) through fresh, relevant, community-driven best practices.

01-10. Explain what Open Service Management is, its characteristics and structure

Module 1: Service management and Open Service Management

Figure 01-10.1 Open Service Management

26

So now that we have a feel for what service management is, what is Open Service Management? Open Service Management, or O-S-M, pronounced, "awesome", s brought to you by the OSM Alliance, www.osmalliance.org, a non-profit whose mission it is to promote fresh, community-driven open service management guidance that is free for individuals and end user organizations to use and share and contribute to.

The general idea is that practitioners are better served through content they can contribute to as well as benefit from, content that is more frequently updated and openly curated. Without a framework with up-to-date best practices, it is incumbent on each IT pro and organization to interpret older best practice guidance for application to their current situation, or to compile best practices from a myriad of emerging best practices, with 100% certainty that what the "roll" will not match what others do. Snapping to an open framework you can contribute to is a better bet.

OSM is shared under a Creative Commons, Attribution-NonCommercial-ShareAlike, "some rights reserved" license, which lets individuals and organizations remix, tweak, and build upon it non-commercially. Licensing is required for commercial use. Commercial rights are reserved for the sole purpose of funding the OSMA's operation, and OSM's development and promotion. Sources of funding for the OSMA include exam fees, Authorized Training Organization (ATO) fees, and Authorized Consulting Organization (ACO) fees, all of which are nominal, as a source of funding for the OSMA's development, promotion and operation.

The OSM is also funded through OSMA consortium member fees. Consortium member organizations have the opportunity to shape the direction of OSM content. The OSMA is also supported by the donations of its members, both in terms of monetary donations, which are tax-deductible, and their contributions of the time and subject matter expertise to contribute to improving OSM to the benefit of our community. To find out more about becoming an OSMA ATO, ACO or consortium member, contact the OSMA at info@osmalliance.org.

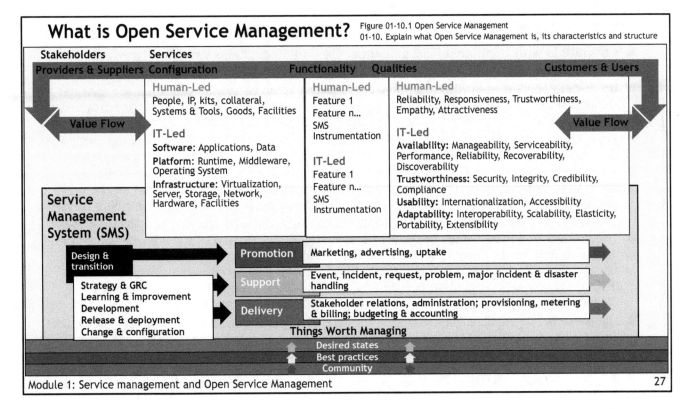

This is the "all up" figure for Open Service Management. Let's dig into it.

First, note that everything above the line that says, "Things worth managing" are the things worth managing in service management: Stakeholders, value and its flow, services, and the SMS.

1. First we have stakeholders; as you see on the left and right of the diagram, these are providers and suppliers, and customers and users; note two differences from traditional ITSM guidance: providers are considered part of the key stakeholder group

2. Next we have value flow; note here another difference from traditional guidance: value is show to flow multidirectionally, to all stakeholders, as this is needed to sustain a service

3. Then we have services; different from traditional guidance, these are subdivided into Human-Led and IT-Led services, and further apportioned into configuration (parts), functionality (features) and qualities (non-functional characteristics).

4. Lastly, we have the SMS, the mechanisms used to continuously and sustainably provide value to stakeholders over time and changing circumstances. The SMS is partitioned into design & transition, promotion, support and delivery, which are not phases and processes, but sets of "things worth managing", each of which has a desired state to achieve and maintain.

At the bottom of the figure you see that the community produces best practices for achieving and maintain desired states for everything above the line, which are things worth managing

> ## What are Open Service Management's characteristics?
> - **Compatible with traditional ITSM guidance.** Like ITIL, etc. OSM is strongly aligned with the ISO/IEC 20000 standard for service management, sharing common terminology and concepts.
> - **Up-to-date** – published in 2018 (traditional ITSM guidance can be over 10 years old, and a lot has changed since then, both in terms of the nature of the target environment that service providers manage, and the best practices—what they actually do that works—that suit current environments.
> - **Free, open, and agile**, with a mechanism to keep it fresh and relevant – OSM is licensed under a Creative Commons Attribution-NonCommercial-ShareAlike "some rights reserved" license; this license lets you remix, tweak, and build upon OSM non-commercially, as long as you credit the OSMA and license you new creations under the identical terms. The intention is that all end-users get OSM for free, and to use it for free, and this freedom give them a reason to contribute to OSM. OSM features an open source pull-style mechanism so all IT pros and organizations can contribute.
> - **The product of a sustainable not-for-profit.** Some rights for commercial use are reserved on the license so to provide a revenue source for the non-profit that owns OSM, the Open Service Management Alliance. The reserved rights include rights for Registered Training Organizations (RTOs), Registered Trainers (RTs), Registered Consulting Organizations (RCOs), and Registered Product Organizations (RPOs) to deliver training, consulting and products aligned to OSM
> - **Light-weight.** One small volume makes up the core OSM Foundation body of knowledge (BOK); OSM Foundation, a compilation of current best practices from, e.g., from Agile, DevOps, and Lean, suited for today's enterprise, hybrid and cloud environments
>
> 01-10. Explain what Open Service Management is, its characteristics and structure
>
> Module 1: Service management and Open Service Management   28

Open Service Management (OSM) is a new and dramatically different set of best practice guidance for service management.

The first thing you should know is that it is strongly aligned to ISO 20000, and as traditional ITSM guidance is also, strongly aligned to ISO 20000, it is compatible with traditional ITSM guidance. Said another way, it can be used alongside traditional ITSM guidance, and understood easily alongside it, as it shares some coming framing and terminology.

So why bother with it if traditional ITSM guidance exists that is commonly referenced?

Well, for one thing, is it's open, so it's worth contributing to and benefitting from. You have the opportunity to make it better. And as it is open and community-driven, we are betting that we can create fresher content at a more agile clip than closed, traditional guidance.

The OSM training and certification path

OSM Foundation training and certification

14 contact hours of training

1-hour, 40-question exam

65% (26 of 40) to pass

Available January 2018

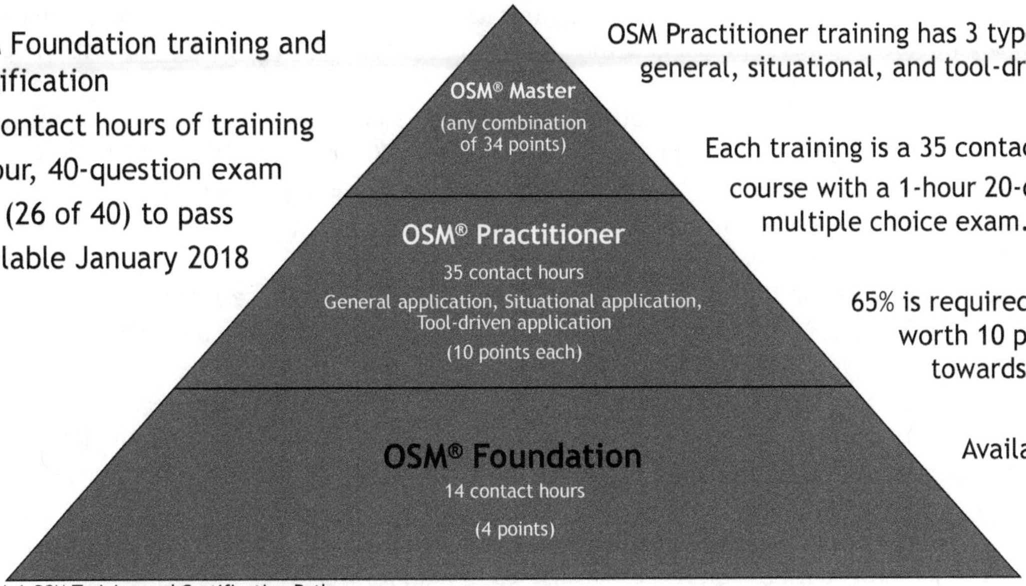

OSM Practitioner training has 3 types: general, situational, and tool-driven.

Each training is a 35 contact-hour course with a 1-hour 20-question multiple choice exam.

65% is required to pass worth 10 points towards Master.

Availability TBA

OSM® Master
(any combination
of 34 points)

OSM® Practitioner
35 contact hours
General application, Situational application,
Tool-driven application
(10 points each)

OSM® Foundation
14 contact hours
(4 points)

Figure 01-11.1 OSM Training and Certification Path
01-11. Explain the training and certification path for OSM (non-examinable)

Module 1: Service management and Open Service Management

The OSM Foundation training and exam have been available since January 1, 2018. You can find the syllabus and sample exams for this training under, "Content" on osmalliance.org.

OSM Foundation training is 14 contact hours. The only prerequisite is that you must take the training from an Authorized Training Organization. This is to help fund the non-profit.

The exam is similar to traditional ITSM exams—one hour, 40 questions, 26 out of 40 to pass, multiple choice. Exam are offered by the OSM Authorized Examination Institute (AEI), acquiros.com.

OSM practitioner courses are under development. The idea here with practitioner is to provide content that helps you apply what you learned in foundation in three ways:

1) generically (practitioner—general application), or in

2) specific situations, like small shops or regulated environments (practitioner—situational application), or with specific

3) Tooling (practitioner—tooling application).

If you would like to contribute to these bodies of knowledge, please contact the OSMA at info@osmalliance.org.

Module Summary (1 of 2)

- Service management is a set of specialized organizational capabilities in the form of a service management system (SMS) for managing the resources required to provide value to customers and users in the form of services
- Things worth managing are the small set of critical things—stakeholders, value flow, services, and the SMS—for which a service provider must aim to achieve and sustainably maintain desired states
- There are four key categories of things worth managing in service management:
    1. **Stakeholders**, those who have a "stake" in service management desired states: customers, users, providers, and suppliers
    2. **Value flow**, which in the multidirectional stream of value through services to stakeholders
    3. **Services**, means for providers to deliver value to customers and users by facilitating desired states customers and users want to achieve
    4. **Service management system (SMS)**, a set of interrelated or interacting mechanisms that service providers use to direct and control their service management activities.
- A desired state is a state of "goodness" to achieve and maintain, sustainably, for a thing worth managing. It is the continuous "end in mind" for things worth managing.
- Best practices are things people are actually commonly doing now that work well when aiming to achieve and maintain desired states / desired states for things worthy of managing

Module Summary (2 of 2)

- Open Service Management (OSM) is a new and dramatically different set of best practice guidance for service management that is free and open, with a mechanism to keep it fresh and relevant, light-weight, and with two editions, enterprise and cloud

- The basis structure of OSM consists of "things worth managing" for service providers, in three key categories—stakeholders, value flow, services, and the SMS, for each thing worth managing, OSM identifies a desired states that the provider must achieve and continuously maintain in order to sustainably provide value to customers and users; lastly, it provides best practices—things IT pros are actually doing now that work—for achieving desired states for things worth managing.

- OSM Foundation training is 14 contact hour with a 40 question 1-hour simple multiple choice exam with 65% required to pass, worth 4 points towards master certification

- OSM Practitioner training is of 3 types—general application, situational, and tool-driven; each practitioner each are 35 contact-hour courses with a 1-hour 20-question multiple choice scenario-based exam with 65% required to pass; each is worth 10 points towards masters

- OSM Masters certification is achieved by any combination of 34 points

Module 2
Stakeholder concepts, desired states and practices (60m)

The OSM® Alliance
Open community-driven best practices

This is the second module of the course, where we focus on stakeholders.

OSM Foundation Course Agenda

Module 1: Service management and Open Service Management	**60m**
Module 2: Stakeholder concepts, desired states and practices	**60m**
Module 3: Value and value flow concepts, desired states and practices	**60m**
Module 4: Service concepts, desired states and practices	**240m**
Module 5: SMS concepts, desired states and practices	**300m**
Module 6: Summary and exam preparation	**120m**

The minimum contact hours for this course are 14, including 2 hours for sample examination revision.

In this module, we will introduce the concept of stakeholders, and their desired states (that is, what "good looks like" for the state of stakeholders, the completion of the statement, "performance is effective when…" for stakeholders.

We'll also cover best practices for achieving and maintain desired states for stakeholders, sustainably.

So let's get started!

Module Objectives

The purpose of this unit is to help you list and describe the stakeholders of service management, their desired states, and best practices for achieving those desired states, including the following stakeholders:

- 02-01. Stakeholders – Overall definition, desired state, best practices
- 02-02. Stakeholders - Customers – Definition, desired state, best practices
- 02-03. Stakeholders - Users – Definition, desired state, best practices
- 02-04. Stakeholders - Providers – Definition, desired state, best practices
- 02-05. Stakeholders - Suppliers – Definition, desired state, best practices

The recommended study period for this unit is minimum 60 minutes or 1 hour.

What specific objectives do you have for these topics?

Stakeholders - the four main stakeholders of service management

Stakeholder

Those with a vested interest in continual value flowing through services to stakeholders, including customers, users, the provider itself, and suppliers, over time and through changing stakeholder needs and technology possibilities.

■ Stakeholders are one of four key categories of things worth managing in service management; there are four primary stakeholder in service management:

Customers	Users	Providers	Suppliers
Those who pay for services and are the primary judge of their value; may be internal (a person or group to the same business as the provider) or external (a person or business separate from the provider)	Those who use the service, but do not pay for it; users may be internal or external to a service provider	Entities that deliver services to customers and users	Third parties who supply providers with services that are required to deliver provider services

Figure 02-01.1 Four primary stakeholders in service management
02-01. Stakeholders - Overall definition, desired state, best practices

Module 2: Stakeholder concepts, desired states and practices   35

Stakeholders are one of the four key "things worth managing" defined by OSM. Stakeholders are people or organizations with a vested interest in their being continual value from a provider's services over time and through changes. In OSM, stakeholders include customers, users, the provider itself, and suppliers.

This is different from traditional ITSM guidance, which does not include the provider on the list of key stakeholders in service management. In OSM, stakeholders are defined as follows:

1. Providers are people or organizations that manage and deliver services to one or more internal or external customers.

2. Customers are people or organizations that pay for and receive services and are the primary judge of their value; they may be internal (a person or group in the same business as the provider) or external (a person or business in a separate legal entity from the provider).

3. Users are people or organizations who use services, including people who use the service but do not pay for it, and customers who pay for the service, in their role as a user of the service; they ma be internal or external to the provider, and

4. Suppliers are people or organizations that are 3rd parties, external to providers (not part of the provider's legal entity) who supply providers with goods or services needed to deliver services to the provider's customers and users.

Stakeholders - desired states

Performance is effective when...

- We know who all our stakeholders are, and what their stake in our services is, and we use this information to systematically maintain productive relationships and tailored communications with all key stakeholders, so there are no or few negative surprises
- We continuously listen to stakeholders and use that listening to "get better reality"—make our stakeholder relationships, service, and service management system better—and "get better perception" - set stakeholder expectations and manage stakeholder perceptions systematically
- When asked, all stakeholders indicate they are pleased with the value they are getting from our services, or are confident, if there is a gap, that we are working to close it and capable of doing so

Stakeholders
Customers
Users
Providers
Suppliers

02-01. Stakeholders - Overall definition, desired state, best practices
Module 2: Stakeholder concepts, desired states and practices

36

Here are some desired states for stakeholders overall. This is not meant to be a "forever"—complete, perfect, and unchanging. Instead, it's meant to be a starter list from which you can identify what effective performance looks like for you relative to stakeholders, the desired state you aim to keep them in.

One key thing to remember here is that unidentified stakeholders (and therefore, stakeholders where the relationship and value have no chance of being systematically kept in a desired state) represent a risk to the service provider.

SO be sure to put mechanisms in place to ensure customers, users, members of the provider organization itself, and suppliers are known, and kept in a desired state of "goodness".

A lot of this can be accomplished through listening posts and feedback loops.

Stakeholders - practices

- <u>Stakeholder analysis / mapping / management</u>
- <u>Stakeholder satisficing</u>
- <u>Moments of Truth</u>
- <u>Getting better reality (Guy Kawasaki)</u>
- <u>Tailored, crisp communication (Mary Munter)</u>
- <u>Feedback loops</u>
- <u>Satisfaction = Perception – Expectation (David Maister)</u>
- <u>Conditions of Satisfaction (Winograd & Flores)</u>
- <u>Customer experience management (CEM)</u>
- <u>User Experience</u>

> Stakeholders
> Customers
> Users
> Providers
> Suppliers

One example of a practice: Satisfaction = Perception – Expectation, *(from "Managing the Professional Service Firm", pg. 71, by David Maister)*

One of the most simple laws of service delivery is that if a stakeholders expectations exceed their perceptions, they will be dissatisfied. This is especially significant because a client's perceptions and expectations may not necessarily reflect reality. Therefore, the achievement of stakeholder satisfaction requires the management of both stakeholder expectation and perception.

Customers

Persons or organizations that pay for and receive services and are the primary judge of their value; they may be internal (persons or groups within the same business as the provider) or external (persons or businesses or business units within a separate legal entity from the provider. Customers may also be users of services.

Stakeholders
→ Customers
Users
Providers
Suppliers

02-02. Stakeholders - Customers - Definition, desired state, best practices
Module 2: Stakeholder concepts, desired states and practices

38

Customers are individuals, organizations, or units within organizations that pay for services, may specify requirements for new services and service improvement, and who are the primary judge of the value of services; they may be internal to the service provider (a person or group within the same legal entity as the provider) or external (a person or business or business unit within a separate legal entity from the provider); customers who pay for services may also be users of those services.

There is a distinct difference between a customer, who pays the bills, and a user of a service, who does not. If you have children, and they have mobile service on your plan, you are the customer, and they are the users. You and your children are in distinctly different roles relative to the service, and the provider, to be successful, has to recognize this, and have mechanisms in place to keep each of you in a desired state of happy with their service.

Having said this, it is also true that being a customer is a role; so customers can also be users of services—think about it; using the example above, if you have children, and pay for their mobile service—you are the customer; but you also use it yourself, which makes you a user, too.

Customers – desired state

Performance is effective when...

- We have a good working relationship with "pay the bills" customers by seeking to both understand and shape customer needs, and by working to ensure those needs are met through the right set of services, which are informed by both changing customer needs and technology possibilities
- We automate our relationship with customers, and that automation facilitates satisfaction (versus being a blocker to it), e.g., through a portal where they can administer their subscription, make requests, log incidents, etc.
- We know which services are used by what customers, and to what extent
- Conflicts with and among customers are rare, but when they do occur, it is easy for the customer to escalate, we handle them promptly and effectively
- We take the time to understand the value of our services from the customers' perspective, and not just our own, and use that understanding to manage both the reality and perception of our services
- Customers say they are highly satisfied with, and continue to buy services, refer us, rarely complain, are engaged, and when asked, speak positively about our services, and say they meet their requirements and compare well against alternatives, as business conditions and technology and the value equation changes over time

Stakeholders
Customers
Users
Providers
Suppliers

In the end we are aiming to keep customers in a known, good state. This list should be a good start, but what is that known, good state for you and your customers? What does performance look like when it is effective? Where are you, relative to that state—grey, yellow, red, or green? What are you doing to get and stay "green"?

- Customer experience management
- Business relationship management
- Self-service portals
- Public status pages
- Multi-channel support
- SERVQUAL
- Service level management
- Service catalog management
- Service portfolio management
- Service reporting
- Service reviews

- Service improvements
- Service retirement

Stakeholders
Customers
Users
Providers
Suppliers

One example of a practice to keep customers in a known, good state is that of public status pages (or transparent uptime).

A public status page give you a place to transparently show information about the availability and performance of your services. You can post announcements, update current issues, and allow people to opt-in to notifications.

In his blog on transparentuptime.com , Lenny Rachitsky gives four points he recommends as prerequisites for doing public status pages

1. Admit failure. Really own it.

2. Sound like a human.

3. Have a communication channel.

4. Be authentic.

Making your current and past performance relative to uptime public is a clear demonstration to customers that you take uptime seriously.

Users – definition and types: end, super, and admin

Users

Persons or organizations who use services, including those who use the service but do not pay for it, and customers who pay for the service, in their role as a user of the service; users may be internal or external to the provider's organization, internal and external users, respectively.

There are different types of users; each type must be recognized and systematically "loved" to ensure satisfaction; user types include:

- End users (whose role is limited to using the service and asking for support or service items they are entitled to)
- Super users (users who have some level of administrative privilege for other users, e.g., to assign roles within a project system)
- Admin users (who administer a subscription / service for others, for example, Office365 admin users)

Stakeholders
Customers
Users
Providers
Suppliers

The user literally uses—interact with your service, product, website or app. You want to make sure your services are easy to use, and actually help users do a unit of work using them.

Also, it may be useful to differentiate between the different types of users, to ensure the mechanisms you have in place to ensure they are in a "good" state of happiness are differentiated enough based on their particular roles, needs and expectations. The example given here is that of end users, super users, and administrative users; this may be a useful set of user types for your purposes, or you may want to differentiate further if it suits your user base and purposes.

Users – desired states

Performance is effective when...

- We have a good relationship with users of our services; we seek to understand and shape user needs, and meet those needs through our services.
- Users say they are highly satisfied with, and continue to use services, refer us to others for services, rarely complain, are engaged, and when asked, speak positively about our services
- We take the time to understand the value of our services from the users' perspective, and not just our own, and use that understanding to manage both the reality and perception of our services
- We automate our relationship with users, and that automation facilitates satisfaction (versus being a blocker to it), e.g., through a portal where they can make requests, log incidents, etc.
- Users say our services help them meet their objectives, and that they are getting higher value from our services as compared to alternatives, including doing it themselves, at a lower effort and risk

Stakeholders
Customers
Users
Providers
Suppliers

02-04. List and describe desired states associated with users

Module 2: Stakeholder concepts, desired states and practices   42

So theses are some ideas for what good performance looks like when users are kept in a desired state.

Some additional indicators of effective performance include:

- Users see us as providing the right mix of services, each with the right mix of features and attractive quality

- Users are happy with the features (including the pace of introduction of new features), performance, service levels and support for our services

- Users understand the terms of our service and their responsibilities as users of our service(s), and

- This feedback is steady over changing services, time, and circumstance, including fluctuations in user demand for our services, and changes in the user environment

Users – practices

- <u>User experience management (UX)</u>
- <u>Self-service portals</u>
- <u>Service desk</u>
- <u>Public status pages</u>
- <u>Multi-channel support</u>
- <u>User satisfaction surveys</u>
- <u>User training</u>
- <u>User story</u>
- <u>Continuous Delivery</u>

Stakeholders
Customers
Users
Providers
Suppliers

These are some practices you can use to drive towards a desired state for user of your services.

One example of these is: Continuous Delivery, or CD, which can help to drive towards the desired state for users because it enables users to start using new functionality and qualities quickly to realize value and provide feedback sooner.

Continuous Delivery, or CD, is about automated implementation of the application build, deploy, test and release processes meant to ensure performance (fast delivery) and conformance (to requirements), enabling users to start using new functionality and qualities quickly to realize value and provide feedback sooner.

Stakeholders - Providers - definition

Provider

Person or organization that manages and delivers services to one or more internal or external customer and users.

- There are several types of provider staff, each of which must be recognized and systematically related to, including:
  - Developers
  - Infrastructure Engineers
  - Support staff
  - Delivery staff
  - Managers

Stakeholders
Customers
Users
Providers
Suppliers

For sustained success at service provision, it is imperative that team members within the provider are kept in a desired state; for example, they should be satisfied with their work.

www.thebalance.com/improve-employee-satisfaction-1917572 indicates this is a function of respectful treatment, fair compensation, benefits, job security, trust, opportunities to use skills and abilities, financial stability of the organization, the employee's relationship with their immediate manager, feeling safe in the work environment, and the employee's immediate manager's respect for their ideas.

Performance is effective when...

- We maintain a positive, effective and sustainable working relationship among provider staff
- We get good value out of our providing services, and customers and users and suppliers do as well
- We automate our relationships internally among different parts of our provider organization, and that automation facilitates satisfaction (versus being a blocker to it), e.g., through a portal where provider staff can make requests, log incidents, etc.
- We take the time to understand the value of our services from the suppliers' perspective, and not just our own, and use that understanding to manage both the reality and perception of our services
- We hire the right people in the first place, and they happy, healthy, and productive, with a service orientation, continuously learning and innovating, rarely complain, are engaged, and when asked, speak positively about their job and our organization

Stakeholders
Customers
Users
Providers
Suppliers

02-04. Stakeholders - Providers - Definition, desired states and practices

Module 2: Stakeholder concepts, desired states and practices

45

Here are some indicators that the provider is in a good, known state. Some additional indicators of effective performance for providers include:

- Our people have the right skills, knowledge, and mindset to succeed, and are good at managing, facilitating meetings, communicating clearly, negotiating and problem solving

- We make time for continuous learning, development, and improvement, about things

  - to achieve and maintain the desired state of things worth managing, and things

  - required to understand what makes customers and users successful and how our services contribute to that, including new technologies that provide advantages

- We have good financial performance

- We have the resources (money, infrastructure, applications, information, and people), capabilities, and authority to succeed

- We have clear and shared goals and objectives

- We have capabilities that are hard for competitors or customers to duplicate

- We manage both the reality and perception of our services

- We have good financial performance

- We have satisfied customers, and

- Our people are aware of our business objectives and those objectives inform what they do

SH

- <u>Hiring the right people in the first place</u>
- <u>Employee engagement</u>

M1

- <u>OLAs</u>
- <u>RACI</u>
- <u>Roles: Process owner, manager, practitioner; service owner</u>

M2

- <u>CAMS</u>

Stakeholders
Customers
Users
Providers
Suppliers

Best practices must be aimed at the reality and perception of the employee on the following parameters: respectful treatment, fair compensation, benefits, job security, trust, opportunities to use skills and abilities, financial stability of the organization, the employee's relationship with their immediate manager, feeling safe in the work environment, and the employee's immediate manager's respect for their ideas.

Stakeholders - Providers - practices

M2
- Affinity (DevOps)
- Blameless Culture (DevOps)
- Collaboration (DevOps)
- Communication, Osmotic (DevOps)
- Self-organizing Teams
- Product Owner (in Agile Scrum)
- Roles: Service Owner, Scrum Master, Scrum Team, embedded teams
- Scaling (DevOps/Agile Scrum)

Affinity (DevOps)
Shared goals, empathy and learning among different groups of people

Blameless Culture (DevOps)
Culture where people are not afraid to fail or experiment because the organizational reaction to them doing so is for all to learn, not to place blame

Collaboration (DevOps)
Targeting a specific outcome through supporting interactions and the input of multiple people

Communication, Osmotic (DevOps)
The idea that if people are in the same room, information will drift through the background to be picked up informally, as by osmosis.

Self-Organizing Teams (Agile Scrum)
Teams that choose how best to accomplish their work (including deciding direction, what work will be done, by whom, and how, and what done and good looks like, monitoring and managing work, and executing the work), rather than being directed by others outside the team

Stakeholders
Customers
Users
Providers
Suppliers

Source: alistair.cockburn.us/Osmotic+communication
02-04. Stakeholders - Providers - Definition, desired states and practices

One example of a practice to keep provider staff in a desired state is embedded teams.

The DevOps practice of embedded teams can help bring down barriers between disparate, siloed teams. An embedded product team consists of all the people and skills needed to independently take product all the way from requirements to delivery.

In DevOps, which is Dev + Ops, the place to start is by embedding Dev representatives in Ops teams, and vice versa.

Stakeholders - Suppliers - Definition

Supplier

Person or organization that is a 3rd party, external to providers (not part of the provider's legal entity) who sign contracts committing themselves to supplying providers with goods or services the provider needs to deliver services to the provider's customers and users.

- There are many types of suppliers, each of which must be related to systematically, including:
- Commodity, tactical and strategic suppliers
- Goods suppliers and service suppliers
- Transactional suppliers and subscription-based suppliers

Stakeholders
Customers
Users
Providers
Suppliers

02-05. Stakeholders - Suppliers - Definition, desired states and practices
Module 2: Stakeholder concepts, desired states and practices

48

Examples of suppliers include:

- Hardware vendors
- System software vendors
- Network vendors
- Application vendors
- DBMS vendors
- Goods vendors
- Services vendors
- Hosting companies

Stakeholders - Suppliers - desired state

Performance is effective when...
- We get good value out of suppliers, and suppliers get good value out of supplying services to us
- We know who our suppliers are, including contacts, contracts, and historical performance, and we regularly review contracts for suitability / performance
- Our contracts have the right legal language to protect us and our relationship with the supplier
- Contractual disputes are rare, but when they do occur, we have an effective process for resolving them
- We automate our relationship with suppliers, and that automation facilitates satisfaction (versus being a blocker to it), e.g., through a portal where they can view and report performance against UCs, etc.
- We do not have to deal with the day to day details of the costs or risks associated with the services suppliers provide us
- We have the right balance between what we do internally and what is outsourced to suppliers

Stakeholders
Customers
Users
Providers
Suppliers

Other indicators of effective performance include:
- Suppliers consistently meet or exceed our business needs and contractual commitments for the goods and services we depend on from them
- We have a good relationship with suppliers, and we proactively manage their performance
- Contract terms with all suppliers are good
- We are not using expensive / higher risk / custom supplier services and goods where cheap / low risk / commodity services and goods will do
- Supplier staff know our key business desired states and how their services support them
- Supplier staff are aware of our patterns of activity, and carry out their activities with this knowledge in mind, avoiding issues
- Supplier transitions go smoothly
- When we manage stakeholder relations, services and the SMS, we take into consideration the suppliers' patterns of business activity
- We have an effective an efficient process for renewing and terminating contracts
- We ensure all supplier contracts support business needs and provide value for money
- We sign up for the right contracts in the first place, with the right commitments in them
- Suppliers used and the goods and services we acquire from them are aligned to our supplier policies, and are compliant with applicable regulatory requirements
- We categorize our suppliers from commodity to strategic, based on the value and risk they present to us; we use this information effectively to apportion our time to suppliers in managing their performance appropriately, in proportion to their importance to us
- Suppliers consistently meet the commitments incorporated in our contracts with them
- While finance and legal may be involved in ensuring supplier contracts are sound, we take ownership for making sure contracts support services and service level and manage the performance of suppliers to their contractual commitments.

Stakeholders - Suppliers - practices

- **Supplier management** – contacts, contracts, performance
- **Vendor value management**
- **Contract management**
- **Underpinning contracts**
- **Supplier categorization**
- **Cloud cost containment**

Stakeholders
Customers
Users
Providers
Suppliers

02-05. Stakeholders - Suppliers - Definition, desired states and practices
Module 2: Stakeholder concepts, desired states and practices

50

An example: Supplier categorization. The Kraljic Portfolio Purchasing Model was created by Peter Kraljic and first appeared in the Harvard Business Review in 1983.

The model involves four steps:

1. Purchase classification

2. Market analysis

3. Strategic positioning

4. Action planning

You can use this model to classify your suppliers and apportion your time during the course of the year according to whether they are strategic suppliers, commodity suppliers, or somewhere in between.

Module Summary

- Stakeholders, along with services and the service management system, are one of the four categories of things worth managing in OSM

- There are four primary stakeholders in service management: 1) customers, 2) users, 3) the provider, and 4) suppliers

- As providers, we seek to continuously provide value to stakeholders through services over time and through changes by achieving and sustainably maintaining desired state for things worth managing through best practices

- For services to be sustainable, value must flow multidirectionally to all stakeholders

- There are desired states for each stakeholder; a number of shared, mode 1, and mode 2 practices exist to help providers achieve and maintain those states.

Stakeholders
Customers
Users
Providers
Suppliers

Module 3
Value and value flow concepts, desired states and practices (60m)

The OSM® Alliance
Open community-driven best practices

This is the third module module of the course, where we focus on value and value flow.

OSM Foundation Course Agenda

Module 1: Service management and Open Service Management	60m
Module 2: Stakeholder concepts, desired states and practices	60m
Module 3: Value and value flow concepts, desired states and practices	60m
Module 4: Service concepts, desired states and practices	240m
Module 5: SMS concepts, desired states and practices	300m
Module 6: Summary and exam preparation	120m

In this module, we'll cover concepts of value and value flow.

Module Objectives

The purpose of this unit is to help you describe value and value flow in service management, its desired state, and best practices for achieving that desired state, including:

03-01. Define what value is, and its characteristics
03-02. Describe value flow in relation to value
03-03. Define value stream, and describe value stream mapping
03-04. Describe Lean, its relation to value, and the five lean principles
03-05. Describe Agile, its relation to value, and the four agile value statements
03-06. Value flow – Overall definition, desired state, best practices

The recommended study period for this unit is minimum 60 minutes, or 1 hour.

What specific objectives do you have for these topics?

What is value? And where does it reside?

Value (of a service)

The worth, in the mind of stakeholders, of a product or service, which is a function of the extent the stakeholder appraises it to fulfill a need or desire. Key components of value of a service include the price (for a customer) or profit (for a provider or supplier), functionality (features) and qualities of the service. Other components of value include enhanced market share, competitive advantage, capability, compliance, safety, reputation, goodwill, and learning. Value must flow multidirectionally, to and from all stakeholders, for services to be sustainable.

Since value is in the mind of stakeholders, as providers we must influence both the reality of our services and their perception by stakeholders

Source: Adapted from http://www.ame.org/sites/default/files/query_archive_docs/LeanGlossary_01_08_1.pdf
03-01. Define what value is, and its characteristics

Module 3: Value and value flow concepts, desired states and practices

55

Here again you should see some similarities and differences between OSM's conception of value, and that of tradition ITSM guidance.

One key difference is that while in OSM, the pay-the-bills customer is the PRIMARY arbiter of value (as without them, you have no business), all stakeholders have a stake in value, and a perspective that must be honored for sustainable success.

Another key difference is that in OSM, value must flow multidirectionally, that is, in all directions, among all stakeholders (customers, users, the provider, and suppliers) for a service to be sustainable.

When is a service valuable to stakeholders?

- For customers, a service is of value when it is attractive as compared to alternatives, in terms of price, risks, and effort required (including the need to attend to details), and where what is gained exceeds what is lost by using it, and where it helps achieve outcomes whose worth exceeds the service cost
- For users, a service is of value when it helps them achieve a desired outcome without undue effort as compared to alternatives
- For the provider, a service is of value if it results in profits, enhanced market share, competitive advantage, capability, compliance, safety, reputation, goodwill, and learning
- For suppliers, the service is of value if it results in the same outcomes as that of the provider, but for the supplier's portion of the service, and for the supplier's business

Source: Adapted from http://www.ame.org/sites/default/files/query_archive_docs/LeanGlossary_01_08_1.pdf
03-01. Define what value is, and its characteristics
Module 3: Value and value flow concepts, desired states and practices

So you see, value is not the same for each stakeholder, because "what you see depends on where you sit." Each stakeholder has a unique stake, and therefore, each service presents a different value equation for each.

Three other key differences between OSM and more traditional ITSM guidance are:

1.   The emphasis on value flow, in line with agile and lean practices.

2.   Value flow as multi-directional (in more traditional guidance, it flows in one direction, from the provider to the customer)

3.   Value as flowing to all stakeholders, with key stakeholders defined as customers, users, the provider, and suppliers

All of these differences are important, especially in the cloud portion of today' typical hybrid environment. Teams are aligned around build pipelines and are working to make value flow unencumbers by inconsistency, variation and overburden. So they get value flow, and their role in making it happen, and are creating and looking for best practices to help them do it.

What is a value stream, and what is value stream mapping?

Value Stream

The specific activities required to design, order and provide a specific product (or service) from concept to launch, order to delivery, or raw materials to finished products into the hands of the customer.

Value Stream Mapping

The identification of all the specific activities occurring along the value stream, represented pictorially in a value stream map, with an eye toward identifying waste, unevenness, and overburden, as a basis for eliminating them, so as to increase the flow of value.

Value Stream mapping (VSM) is typically described in the following four steps:

1. Identify the target product, product family, or service.
2. Draw (preferably by walking on the shop floor or the virtual flow of information) a current state value stream map (CSVSM).
3. Draw a future state (with waste removed) value stream map (FSVSM).
4. Develop an action plan to work toward the future state condition, and implement the plan.

Source: Adapted from http://www.ame.org/sites/default/files/query_archive_docs/LeanGlossary_01_08_1.pdf

03-03. Define value stream, and describe value stream mapping

Value stream mapping is a critical technique in OSM, because the flow of value through services is critical to success at service management. There are many places to apply VSM. Your build pipeline is a no-brainer to start with if you haven't already.

But keep in mind that value takes many forms and flow multidirectionally in our conception. So, for example, you can apply VSM to feedback loops between, for example, to amplifying feedback loops; for example, you could instantiate tighter feedback loops between BRM, strategy, continual improvement, and sprint planning and retrospectives. In this example, the core value that flows is knowledge—user and customer feedback, changing technology possibilities, changing business conditions—not code, as in a build pipeline. In general, you should seek shorter cycles and smaller, more frequent increments for all four, and that starts with understanding the value stream.

Draw on Lean principles and Agile values to deliver more value, faster

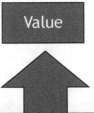

Value
Lean
• Continually improve
• Eliminate unnecessary activities
Agile
• Respond to changing needs
• Shorten release cycles
• Collaborate

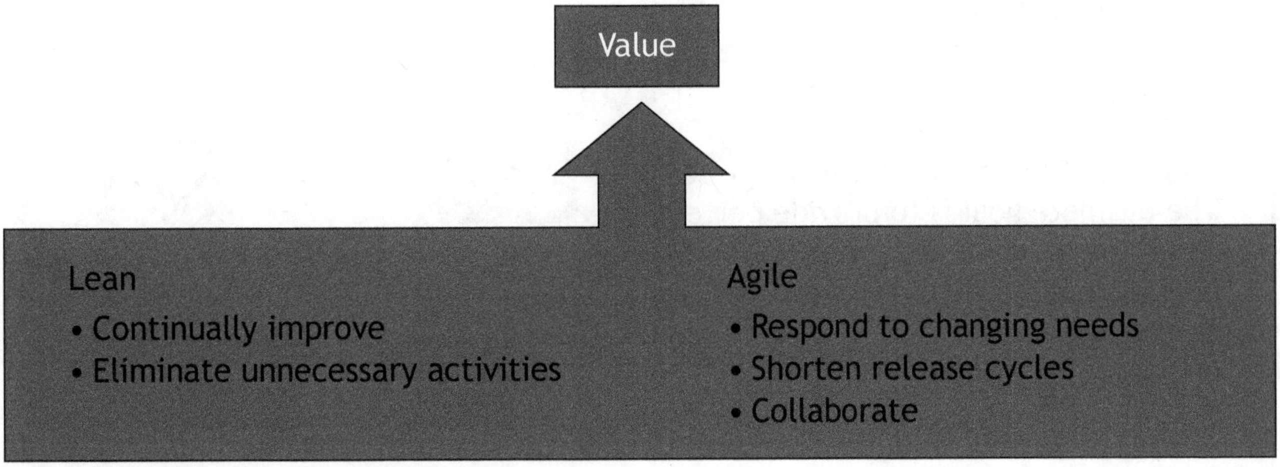

Figure 03-04.1 Lean and Agile and their relation to value
03-04. Describe Lean, its relation to value, and the five lean principles

Module 3: Value and value flow concepts, desired states and practices 59

The five Lean principles are:

1. Identify value
2. Map the value stream
3. Create flow
4. Establish pull
5. Seek perfection

The four Agile values are:

1. Individuals and interactions over processes and tools
2. Working software over comprehensive documentation
3. Customer collaboration over contract negotiation
4. Responding to change over following a plan

Lean, its relation to value, and the five lean principles

Lean

Simply, lean means creating more value for customers with fewer resources. A lean organization understands customer value and focuses its key processes to continuously increase it. The ultimate goal is to provide perfect value to the customer through a perfect value creation process that has zero waste.

Five Lean Principles

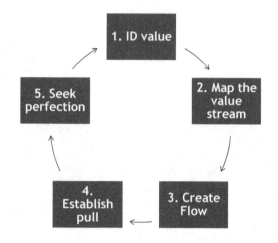

Figure 03-04.1 Five Lean Principles   Source: http://www.ame.org/sites/default/files/query_archive_docs/LeanGlossary_01_08_1.pdf, https://www.lean.org/WhatsLean/
03-04. Describe Lean, its relation to value, and the five lean principles

Module 3: Value and value flow concepts, desired states and practices   60

The five lean principles as cited y ame.org are:

1. Identify Value: When a product or service has been perceived or appraised to fulfill a need or desire--as defined by the customer--the product or service may be said to have value or worth. Components of value may include quality, utility, functionality, capacity, aesthetics, timeliness or availability, price, etc.

2. Map the Value Stream: All the activities (both value-added and non-value added) required within an organization to deliver a specific service; "everything that goes into" creating and delivering the "value to the end-customer.

3. Create Flow: The progressive achievement of tasks and/or information as it proceeds along the value stream, flow challenges us to reorganize the Value Stream to be continuous… "one by one, non-stop".

4. Establish Pull: Principle the no one upstream function or department should produce a good or service until the customer downstream asks for it. Pull is a system of cascading production and delivery instructions from downstream to upstream activities in which nothing is produced by the upstream supplier till the downstream customer signals a need

5. Seek Perfection: A never ending pursuit of the complete elimination of non-value adding waste so that all activities along a value stream create value; perfection challenges us to also create compelling quality ("defect free") while also reducing cost ("lowest cost"). Perfection is Complete elimination of waste (Muda) so that all activities along

Lean systems focus on the parts of the system that add value by eliminating waste everywhere else, whether that be overproduction of some parts, defective products that have to be rebuilt, or time spent waiting on some other part of the system. Waste to be eliminated in these areas can include unnecessary software features, communication delays, slow application response times, and overbearing bureaucratic processes.

Agile

A specific set of values and principles, as expressed in the Agile Manifesto (Beck et al. 2001); An umbrella term used for a group of related approaches to software development based on iterative and incremental development. Scrum is an agile approach to development. See Extreme Programming, Kanban, Scrum.

The Agile Manifesto: The four agile statements of value		
Individuals and interactions	over	processes and tools
Working software	over	comprehensive documentation
Customer collaboration	over	contract negotiation
Responding to change	over	following a plan

Source: agilemanifesto.com

Figure 03-05.1. The Agile Manifesto
03-05. Describe Agile, its relation to value, and the four agile statements of value

Module 3: Value and value flow concepts, desired states and practices

61

"Agile" (with a capital A) refers to a set of frameworks used for development and management of programs and projects intended to make development quicker and smoother, and to create output that is more satisfying for the customer.

So far, you know that an Agile framework is one that uses an adaptive development lifecycle instead of a predictive one. Agility was around back in the 50s, where it was considered a strange and sometimes anarchistic viewpoint; in a time when the traditional method did not even have a name such as Waterfall, because it was the de-facto way of working.

The term "Agile" has become more and more established and was formalized by a group that prepared and signed a statement of values for Agile projects back in 2001, known as the Agile Manifesto.

Items on the right are obviously important but they are a means to an end. Often peple focus on items on the right, thereby spending energy (and time) on the wrong things. Items on the right are the things that really matter to a (business/customer) organization.

Value and value flow - desired state

Performance is effective when...

- Value flows to all stakeholders (customers, users, us (the provider), and suppliers) multidirectionally and unencumbered, with no undue stoppages, scrap, rework or backflows
- When asked, stakeholders indicate they are pleased with the value they get from our services, both in terms of what they get, and the pace at which they get it
- Value flows continually—over time and through changes—from services, to customers and users, and back to the provider and suppliers
- The value from our services compares favorably to alternatives, because we have taken the time to seek out changing business needs and new technology possibilities, and to incorporate that looking and listening into our service
- Uptake of services, features and feature sets is widespread in users and they are getting good value out of them

03-06. Value flow - Overall definition, desired state, best practices

Module 3: Value and value flow concepts, desired states and practices

62

So you see, performance is effective when value flows multidirectionally, to all stakeholders, and is sufficient for the sustainability of the service, as we change it continuously over time to account for changing stakeholder needs and technology possibilities.

What are value and value flow best practices?

- <u>The five lean principles</u>
- <u>The four agile values</u>
- <u>Value attribute tree</u>
- <u>Value stream mapping</u>
- <u>Feedback loops</u>
- <u>Build pipeline</u>
- <u>Continuous integration</u>
- <u>Continuous delivery</u>
- <u>Continuous deployment</u>
- <u>Blue / Green Deployment</u>

An example: blue / green deployment.

Blue/Green Deployment is supplanting rolling upgrades as a method for releasing software code, where two environments, Blue and Green, are initially configured identically, with one active and the other environment inactive. New code is released to the inactive environment, where it is thoroughly tested; Then the active and inactive environments are switched. If problems are discovered after the switch, traffic can be directed back to the inactive configuration that still runs the original version. Once the new code has proven itself in production, the team updates the code in the inactive environment to match the active environment. The process reverses itself when the next software iteration is ready for release.

This technique eliminates downtime due to application deployment, and also reduces risks and dramatically lowers the transaction cost and time required to revert to a prior known, good state; if something unexpected happens with your new version on Green, you can immediately roll back to the last version by switching back to Blue.

Module Summary

- Value is the worth in the mind of stakeholders, of a product or service
- Value must flow to all stakeholders multidirectionally for a service to be sustainable
- Value flow is the multidirectional stream of value delivered through services to all stakeholders, primarily customers, users, the provider, and suppliers
- Value is created and added through design, transition, promotion, support, and delivery activities (which can be fully or partly automated or manual), as a applied to a service and its associates configuration, features, and qualities.
- Lean and Agile principles can help providers deliver value

Module 4
Service concepts, desired states and practices (240m)

This is the fourth module of the course, where we focus on services, their constituent components, desired states, and best practices for driving towards desired states.

OSM Foundation Course Agenda

Module 1: Service management and Open Service Management	60m
Module 2: Stakeholder concepts, desired states and practices	60m
Module 3: Value and value flow concepts, desired states and practices	60m
Module 4: Service concepts, desired states and practices	**240m**
Module 5: SMS concepts, desired states and practices	300m
Module 6: Summary and exam preparation	120m

The minimum contact hours for this course are 14, including 2 hours for sample examination revision.

Here we'll cover service concepts, desired states and practices.

What is a service?

Service

A means for providers to deliver value to stakeholders; customers engage providers for services typically when doing so presents lower cost, effort and risk when compared to doing it themselves or by using alternative services; by doing so, customers aim to focus on ends / desired states versus the details of the means, which the provider handles. Services provide value continuously over time and change when effectively directed and controlled by a service management system

- Services are one of four key categories of things worth managing in service management
- Services can be internal or external, i.e., internal services are services provided to customers and users belonging to the same business entity as the provider; external services are provided to individuals and organizations outside of the provider's business entity

Customer
- Pays for and does not perform services
- Owns costs and risks, but not the details
- Accountable for ends / desired states

Service Provider
- Gets paid to perform services
- Takes on the <u>details</u> of costs and risks
- Responsible for means to achieve desired states

A service is a means of delivering value to customers by facilitating the desired states customers want to achieve, at (what should be) a lower cost and risk than doing it themselves, or other alternatives, and without the need to pay attention to all the details (or means) of delivering the service. The general idea—and you should be familiar with this because this is why you utilize services instead of simply doing it yourself, or using an alternative—is that you can focus on the ends, or outcomes, instead of the means, and because the service presents a better value in the end than doing it yourself or alternatives.

Services must also provide value to users, and value back to the provider and its suppliers, in order for services to be sustainable. Not that in this definition, the value flow is multidirectional; also, the nature of value is not just in new services and features, it is multi-dimensional; for example, value can flow back to the provider from customers and users in the form of feedback on services, feature sets, features, and new feature requests.

What three things are worth managing when it comes to services?

1. **Configuration** (components / what it is made of) that make up the service
2. **Functionality** (features / what it does) the service provides, including telemetry that facilitates direction and control of the service by the Service Management System
3. **Qualities** (characteristics / how it behave) of the service, the non-functional "-ilities" of the service

Each has a desired state to be achieved and maintained through best practices.

IT-Led Services			
Configuration	Functionality	Qualities	
Software • Applications • Data Platform • Runtime • Middleware • Operating System Infrastructure • Virtualization • Server • Storage • Network • Hardware • Facilities	• Feature 1 • Feature n... • SMS Instrumentation	Availability • Manageability • Serviceability • Performance • Reliability • Recoverability • Discoverability Trustworthiness • Security • Integrity • Credibility • Compliance	Usability • Internationalization • Accessibility Adaptability • Interoperability • Scalability • Elasticity • Portability • Extensibility

Human-Led Services		
Configuration	Functionality	Qualities
• People • Service IP, kits, collateral • Systems & tools • Goods • Facilities	• Feature 1 • Feature n... • Feature n... • SMS Instrumentation	• Reliability • Responsiveness • Trustworthiness • Empathy • Attractiveness

Figure 04-01-01. Human-Led and IT-Led things worth managing
04-01-01. Service - Overall definition, desired state, best practices
Module 4: Service concepts, desired states and practices

68

An example, for an IT-Led service (adapted from http://ux.stackexchange.com/questions/47897/what-are-the-differences-between-features-and-components)

for example, take Microsoft Word

Its components:

- spell-checker,

- Page designer,

- Word art etc.

Its features:

- can detect your spelling mistake

- you can customize page-design

- you can draw simple arts, etc.

Its qualities:

- Internationalization

- Accessibility

- Security, etc.

What are the characteristics of services?

- Services can be fully or partially automated or manual
- Services can involve the provision of goods, e.g., replacement keyboard
- Services can be internal or external, i.e., internal services are provided to customers and users belonging to the same business entity as the provider; external services are provided to individuals and organizations outside of the provider's business entity
- End-services are what that the customer recognizes and pay for; they typically include sub-services (e.g., updates, DHCP); services can be broken down into core services (e.g., for mobile phone service, dial tone), optional services (e.g., data plan), and supporting services (e.g. backups, update); services can be stratified in tiers (e.g., gold, silver, bronze level packages)

04-01-01. Service - Overall definition, desired state, best practices
Module 4: Service concepts, desired states and practices

69

Services can have a range of automation. They can involve the provision of goods.

Services can be for internal or external customers and users.

And services can be made up of sub-services, and subdivided into core, optional, and supporting services. They can also feature stratified levels.

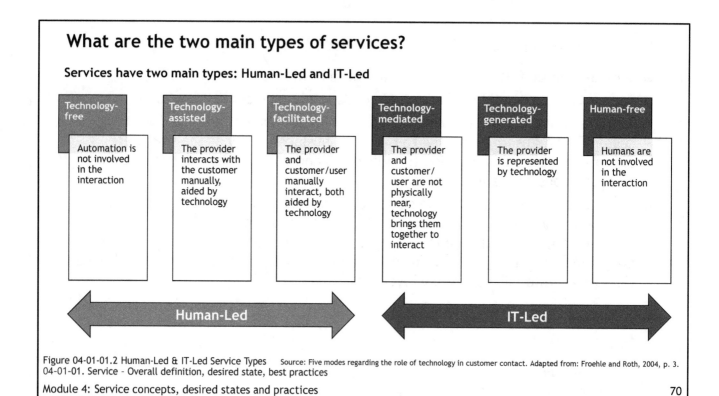

What are the two main types of services?

Services have two main types: Human-Led and IT-Led

Technology-free	Technology-assisted	Technology-facilitated	Technology-mediated	Technology-generated	Human-free
Automation is not involved in the interaction	The provider interacts with the customer manually, aided by technology	The provider and customer/user manually interact, both aided by technology	The provider and customer/user are not physically near, technology brings them together to interact	The provider is represented by technology	Humans are not involved in the interaction

← Human-Led → **← IT-Led →**

Figure 04-01-01.2 Human-Led & IT-Led Service Types   Source: Five modes regarding the role of technology in customer contact. Adapted from: Froehle and Roth, 2004, p. 3.
04-01-01. Service - Overall definition, desired state, best practices

Module 4: Service concepts, desired states and practices   70

There are two types services: 1) Human-Led Services, performed by humans, e.g., moves, adds and changes; technology may be assist / facilitate, but the primary driver is human, and 2) -Led Services, e.g., AD, DNS, etc.; technology is the driver, humans may assist / facilitate.

Services sit along a continuum from technology to technology generated; we've adapted Froehle and Roth's five modes of human and technology interaction model to help illustrate:

Technology-free customer contact For example, the face-to-face contact between a psychological therapist and patient. This type of contact does not require the use of technology; thus, it is the most traditional and direct mode of customer contact.

Technology-assisted customer contact For example, hotel check-in and check-out procedures, transactions conducted over manned bank counters, and passenger check-ins for airline boarding. During these processes, technology (i.e., a computer) is used by the service provider only. However, the customer and service personnel still experience face-to-face contact.

Technology-facilitated customer contact For example, the use of Microsoft PowerPoint by a financial expert to present and discuss financial plans with customers during a conference. Although technology is used by both parties, face-to-face customer contact still occurs.

Technology-mediated customer contact For example, telephonic interaction between service personnel and customers and the provision of professional advice by a consulting company using Videoconference technology. In these contexts, a shared technological platform is used by the service personnel and the customer without face-to-face contact.

Technology-generated customer contact, e.g., ATM withdrawals. In these, customers operate the technology without assistance of service personnel, and face-to-screen contact replaces face-to-face contact. Online shopping is an example of this mode.

We add a 6th one here, Human-free, where technology represents both the provider and the consumer of the service, as in EDI.

What are Human-Led services?

Human-Led Services

Services driven by humans, that may be assisted or facilitated by IT, e.g., PC repair, moves/adds/changes; contrast this with IT-Led services, which are driven by technology, and may be assisted or facilitated by humans, e.g., Office365.

Includes:

- Technology-free services, like face-to-face consultation
- Technology-assisted services, like a service desk rep using a knowledgebase, where only the provider uses technology
- Technology-facilitated services, like virtual instruction, where both providers and consumers use the technology

Human-Led Services		
Configuration	Functionality	Qualities
• People • Service IP, kits, collateral • Systems & tools • Goods • Facilities	• Feature 1 • Feature n... • Feature n... • SMS Instrumentation	• Reliability • Responsiveness • Trustworthiness • Empathy • Attractiveness

Figure 04-01-01.3 Human-Led Services
04-01-01. Service - Overall definition, desired state, best practices

Module 4: Service concepts, desired states and practices

71

Human-Led services include:

- Technology-free customer contact For example, the face-to-face contact between a psychological therapist and patient. This type of contact does not require the use of technology; thus, it is the most traditional and direct mode of customer contact.

- Technology-assisted customer contact For example, hotel check-in and check-out procedures, transactions conducted over manned bank counters, and passenger check-ins for airline boarding. During these processes, technology (i.e., a computer) is used by the service personnel only. However, the customer and service personnel still experience face-to-face contact.

- Technology-facilitated customer contact For example, the use of Microsoft PowerPoint by a financial expert to present and discuss financial plans with customers during a conference. Although technology is used by both parties, face-to-face customer contact still occurs.

Service Type - IT-Led Services - Definition

IT-Led Services

Services driven by information technology that may be assisted or facilitated by humans, e.g., Office365. Contrast this with Human-Led services, which are driven by humans, and may be assisted or facilitated by IT, for example, PC repair, Moves/Adds/Changes.

Includes:

- Technology-mediated
- Technology-generated
- Human-free

IT-Led Services			
Configuration	Functionality	Qualities	
Software • Applications • Data Platform • Runtime • Middleware • Operating System Infrastructure • Virtualization • Server • Storage • Network • Hardware • Facilities	• Feature 1 • Feature n... • SMS Instrumentation	Availability • Manageability • Serviceability • Performance • Reliability • Recoverability • Discoverability Trustworthiness • Security • Integrity • Credibility • Compliance	Usability • Internationalization • Accessibility Adaptability • Interoperability • Scalability • Elasticity • Portability • Extensibility

Figure 04-01-01.4 IT-Led Services
04-01-01. Service - Overall definition, desired state, best practices

IT-Led services include

- Technology-mediated customer contact For example, telephonic interaction between service personnel and customers and the provision of professional advice by a consulting company using Videoconference technology. In these contexts, a shared technological platform is used by the service personnel and the customer without face-to-face contact.

- Technology-generated customer contact For example, ATM withdrawals, vending machine purchases, or coin-operated automatic photograph booth operations. In these processes, customers operate the technology without the assistance of service personnel, and face-to-screen contact replaces face-to-face customer contact. The recent trend of online shopping is considered an example of this mode.

- Human-free, where technology represents both the provider and the consumer of the service, as in EDI.

What is the desired state for services?

Performance is effective when...

- We have the right mix of IT-Led and Human-Led services, with the right level of automation and human activities in each
- We have rooted out of IT-Led services any unnecessary complexity, variation, and dependencies and any other factors that slow down stakeholders, services, and the SMS
- Customers and users are happy with the value they are getting from our services and are sticking with us, because we as time and circumstances change, we listen to how their needs and value equation is changing, and understand and educate them on what new technologies can offer, and bring these two things together to continuously add, change, and remove services so our portfolio is compelling to them
- We have both "better reality" and "better perception" in our services in the eyes of customers and users

04-01-01. Service - Overall definition, desired state, best practices

Some additional indicators of effective performance:

- Customers and users can focus on their "ends-in-mind" because we handle the details of the services we provide, and because we have (and they see we have) the specialized capability to provide the services they value at a lower cost, effort and risk than doing it themselves or as compared to alternatives

- The provider (us) and our suppliers are happy with the value we are getting from providing services over time and through changes

- Services are profitable, over time and across changes, because we add, modify, and remove services continuously to achieve and increase value as time passes and circumstances change

04-01-01. Services – practices

- For services, overall best practices are divided into Human-Led and IT-Led, and within each, practices to achieve and maintain desired states for configuration, functionality, and qualities, as well as shared, mode 1 and mode 2 best practices suitable for all environments, traditional IT, and cloud environments

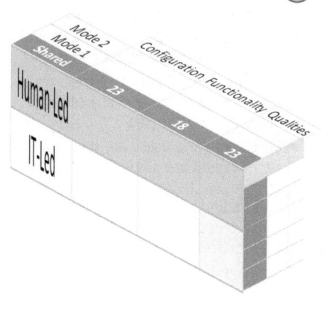

Figure 04-01-01.1 Overall services best practices
04-01-01. Service – Overall definition, desired state, best practices

Module 4: Service concepts, desired states and practices

For services, overall best practices are divided into Human-Led and IT-Led, and within each, practices to achieve and maintain desired states for configuration, functionality, and qualities, as well as shared, mode 1 and mode 2 best practices suitable for all environments, traditional IT, and cloud environments. Configuration is what the service is made up of, it's parts; functionality is its features / what it does; qualities are its characteristics / how it behaves.

Service Configuration

The purpose of this unit is to help you describe the components that make up the configuration of services, their desired states, and best practices for achieving them, including:

04-01-02. Service Configuration

04-01-03. Human-Led Services

04-01-04. IT-Led Services

04-01-05. Software

04-01-06. Applications

04-01-07. Data

04-01-08. Platform

04-01-09. Runtime

04-01-10. Middleware

04-01-11. Operating System

04-01-12. Infrastructure

04-01-13. Virtualization

04-01-14. Server

04-01-15. Storage

04-01-16. Network

04-01-17. Hardware

04-01-18. Facilities

Module 4 Lesson 1

Service configuration definitions, desired states and practices (120m)

75

This is the first lesson in module four, where we focus on definitions, desired states and practices for the configuration portion of services. In OSM, services are made up of three things: configuration (parts), functionality (features), and qualities (non-functional characteristics, like availability).

Service configuration - definition

Configuration

The component parts (what it is made of) of a Human-Led or IT-Led service, where Human-Led service components can include people, IP and service kits, collateral, systems and tools, goods and facilities, and IT-Led services may be composed of software, platforms, and infrastructure; contrast this with service functionality (features / what it does) and qualities (characteristics / how it behaves).

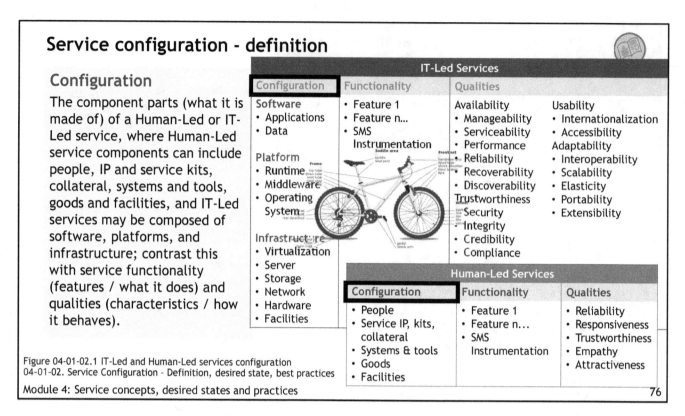

IT-Led Services			
Configuration	**Functionality**	**Qualities**	
Software • Applications • Data Platform • Runtime • Middleware • Operating System Infrastructure • Virtualization • Server • Storage • Network • Hardware • Facilities	• Feature 1 • Feature n... • SMS Instrumentation	Availability • Manageability • Serviceability • Performance • Reliability • Recoverability • Discoverability Trustworthiness • Security • Integrity • Credibility • Compliance	Usability • Internationalization • Accessibility Adaptability • Interoperability • Scalability • Elasticity • Portability • Extensibility

Human-Led Services		
Configuration	**Functionality**	**Qualities**
• People • Service IP, kits, collateral • Systems & tools • Goods • Facilities	• Feature 1 • Feature n... • SMS Instrumentation	• Reliability • Responsiveness • Trustworthiness • Empathy • Attractiveness

Figure 04-01-02.1 IT-Led and Human-Led services configuration
04-01-02. Service Configuration - Definition, desired state, best practices

Module 4: Service concepts, desired states and practices   76

An example may help illustrate: For a bicycle delivery service in a metropolitan area, the configuration would include the couriers, the fleet of bikes, the repair shop, etc. The features would include different size and type packages that could be dropped off. The qualities would include the delivery area, and so on.

Service configuration example: Human-Led service – Office365

Configuration (parts, what it is made of)	Functionality (features, what it does)	Qualities (characteristics, how it behaves)
Word • Spell-checker • Page designer • Word art etc.	**Word** • can detect your spelling mistake • you can customize page-design • you can draw simple arts etc.	**Word** • Internationalization • Accessibility • Security
Excel • Formula editor • Graph editor • Diagram etc.	**Excel** • can detect your formula mistake • you can customize overall sheet design • you can create charts from your data etc.	**Excel** • Internationalization • Accessibility • Security

Source: adapted from http://ux.stackexchange.com/questions/47897/what-are-the-differences-between-features-and-components

Figure 04-01-02.2 IT-Led and Human-Led services configuration

04-01-02. Service Configuration – Definition, desired state, best practices

Module 4: Service concepts, desired states and practices

77

You should see that the configuration, or parts, are distinct from the functionality, or features, and the qualities, or characteristics.

Each contributes (or takes away from) the value of a service, and as such, each needs to be recognized and systematically managed.

Configuration (parts, what it is made of)	Functionality (features, what it does)	Qualities (characteristics, how it behaves)
• Physical concierge area • Spare parts and supplies • Tools, etc.	• Cracked screen repair • Warrantee repair • Accessory sales (USB chords, chargers, mice, keyboards, mouse mats, etc.) • Virus removal • Upgrades, etc.	• Devices serviced • Products carried • Timelines for repair, etc.

Figure 04-01-02.3 Human-Led service configuration example - device bar
04-01-02. Service Configuration - Definition, desired state, best practices

Module 4: Service concepts, desired states and practices   78

You've probably visited the Apple Genius bar or the Microsoft In-Store Answer Desk at one time or another, or perhaps you have concierge-style support locations in you company.

Think about a few specific instances of when this has worked well for you (or not). Were the positives or negatives driven by the configuration, functionality, or qualities of the service?

Human-Led Services	
Configuration	
• People	
• Service IP, kits, collateral	
• Systems & tools	
• Goods	
• Facilities	

Human-Led Services

Services provided primarily by humans, possibly with the assistance of technology, but where the human-to-human interaction is the primary driver. Contrast this with IT-Led Services, where services are provided primary by technology, possibly with the assistance of humans, but where technology-to-technology or technology to human interaction is the primary driver.

Human-Led service components include:

■ People – the roles and individuals and teams that deliver the service and their skills, knowledge and mindset(s)

■ Service IP, kits, collateral—the knowledge and materials teams rely on to deliver services

■ Systems & Tools—for example, a process template for a process engineering engagement

■ Goods—for services where goods are involved, e.g., installing a new physical monitor

■ Facilities—for example, a physical location for concierge-style device support

Examples		
■ Moves, adds and changes	■ New hire setup	■ Consulting
	■ Legal hold	■ Auditing

Figure 04-01-03.1 Human-Led services configuration
04-01-03. Service Configuration – Human-Led Services – Definition, desired state, best practices

For value to be realized by stakeholders of your Human-Led services, you must keep the components of Human-Led services in good repair, so that they contribute to a desired state of being reliable, responsive, trustworthy, empathetic, and with good attractiveness.

You most certainly have been on the receiving end of Human-Led services, like getting your office moved, or using consulting services.

Think about a few specific instances of when this has worked well for you (or not). Were the positives or negatives driven by the configuration, functionality, or qualities of the service?

Human-Led service configuration - desired state

Human-Led Services
Configuration
• People
• Service IP, kits, collateral
• Systems & tools
• Goods
• Facilities

Performance is effective when...

- People—individuals and teams that deliver the service, are the right people with the knowledge, skills and mindset, and are attractive in appearance

- Service IP, kits, collateral—the knowledge and materials teams rely on to deliver services—is up to date, complete, and easy to locate, navigate and use

- Systems & Tools—for example, a process template for a process engineering engagement—exist and are sufficient to ensure successful delivery

- Goods—for services where goods are involved, e.g., installing a new physical monitor—we manage inventory to meet demand, and goods are stored in an organized fashion so they are readily accessible, and not shoddy looking

- Facilities—are attractive in their appearance and strongly support conducting the service

Figure 04-01-03.1 Human-Led service qualities; Source: https://en.wikipedia.org/wiki/SERVQUAL
04-01-03. Service Configuration - Human-Led Services - Definition, desired state, best practices

"Attractive" in this sense does not mean that all our people are models, that all our facilities are slick, etc. It means at a minimum that they are not off-putting, and are suitable for the purposes of the service and its stakeholders. For example, there is one standard of dress you might expect a dollar store employee to have, and quite another for say, a Nordstrom salesperson. And if the dollar store fixtures were too expensive-looking, you might wonder what you're getting for your dollar—similarly, if the fixtures in a Nordstrom's were shoddy or cheap looking, you might be wondering if the t-shirt you're buying is really worth $30.

For value to be realized by stakeholders of your Human-Led services, you must keep the components of Human-Led services in good repair, so that they contribute to a desired state of being reliable, responsive, trustworthy, empathetic, and with good attractiveness. This means keeping the following in good order:

- People—the individuals and teams that deliver the service

- Service IP, kits, collateral—the knowledge and materials teams rely on to deliver services

- Systems & Tools—for example, a process template for a process engineering engagement

- Goods—for services where goods are involved, e.g., installing a new physical monitor

- Facilities—for example, the tangible appearance of a physical service location

Human-Led service configuration – practices

Human-Led Services
Configuration
• People
• Service IP, kits, collateral
• Systems & tools
• Goods
• Facilities

- **SERVQUAL** Valerie Zeithaml and Mary Jo Bitner
- **Service IP, kits, collateral** Thomas E. Lah
- **Systems and tools** Thomas E. Lah
- **Goods** TSIA
- **Facilities** Micah Solomon

There isn't a lot of good writing out there on packaging the parts of Human-Led service kits. One good source is Thomas E. Lah.

IT-Led service configuration - definition

IT-Led Services

IT-Led Services
Configuration
Software • Applications • Data
Platform • Runtime • Middleware • Operating System
Infrastructure • Virtualization • Server • Storage • Network • Hardware • Facilities

IT-Led Services

Services provided primarily by technology, possibly with the assistance of humans, but where technology-to-technology or technology-to-human interaction is the primary driver. Contrast with Human-Led Services, which are provided primarily by humans, possibly with the assistance of technology, but where human-to-human interaction is the primary driver, e.g., AWS, Azure, Office365, Salesforce

Service configuration

The component parts (what it is made of) of a Human-Led or IT-Led service, where Human-Led service components can include people, IP and service kits, collateral, systems and tools, goods and facilities, and IT-Led services may be composed of software, platforms, and infrastructure; contrast this with service functionality (features / what it does) and qualities (characteristics / how it behaves).

An IT-Led services configuration is comprised of three layers: software, platform and infrastructure. For cloud environments, these are all, "as a service"; for traditional environment, they are not.

Figure 04-01-04.1 IT-Led services configuration
04-01-04. Service Configuration - IT-Led Services - Definition, desired state, best practices

IT-Led services are services where IT takes the lead in representing the provider, anything from being an equal partner with humans, up to fully representing the provider to a human, and beyond—where the provider is represented by technology, and the consumer is, too, as in EDI.

IT-Led service configuration - definition

IT-Led Services
Configuration

Software
• Applications
• Data

Platform
• Runtime
• Middleware
• Operating System

Infrastructure
• Virtualization
• Server
• Storage
• Network
• Hardware
• Facilities

Infrastructure, Software, and Platform or IaaS, SaaS and PaaS

The IT-Led Services stack includes infrastructure, platform, and software components; in traditional IT, you are the provider for the entire stack; in the cloud, "as a service" model, you use an external service provider for:

- **SaaS** (Software as a Service) provides you with access to application software on-demand without having to worry about the installation, setup and running of the application, which the service provider handles. You just have to pay for it and use it through some client. Examples: Google Apps, Office365.
- **PaaS** (Platform as a Service), provides you computing platforms which typically includes operating system, programming language execution environment, database, web server etc. Examples: AWS Elastic Beanstalk, Windows Azure, Heroku, Force.com, Google App Engine, Apache Stratos.
- **IaaS** (Infrastructure as a Service) provides you the computing infrastructure, physical or (quite often) virtual machines and other resources like virtual-machine disk image library, block and file-based storage, firewalls, load balancers, IP addresses, virtual local area networks etc. Examples: Amazon EC2, Windows Azure, Rackspace, Google Compute Engine.

Source: http://stackoverflow.com/questions/16820336/what-is-saas-paas-and-iaas-with-examples
04-01-04. Service Configuration - IT-Led Services - Definition, desired state, best practices

In traditional IT, the software / platform / infrastructure stack is provided on-prem, and not "as-a-service". In a cloud environment, each of these layers is provided, "as-a-service".

The typical environment at the time of this writing is a hybrid of traditional IT and cloud, so it is important to recognize and systematically manage the configuration of both.

IT-Led service configuration – desired state

IT-Led Services
Configuration

Software
• Applications
• Data

Platform
• Runtime
• Middleware
• Operating
  System

Infrastructure
• Virtualization
• Server
• Storage
• Network
• Hardware
• Facilities

Performance is effective when…

■ We have the right mix of traditional and cloud-based IT-Led services, or are moving towards that mix at the right pace

■ We know what we have, what we are paying for and using (and we are not paying for stuff we aren't using)

■ We have a balance of flexibility for choice of services for customers and users, but not such a proliferation of options that the work of customers and users is negatively impacted (learning curve, interoperability issues, etc.)

■ Our IT-Led service configuration "adds up" to consistently and sustainably support the levels of service expected by stakeholders, over time and through changing stakeholder needs and new technology possibilities

Knowing what you have so you can manage it sounds like a basic and expected thing, but the reality, with things like growth through acquisition, and the proliferation of people buying cloud services, is that it "ain't necessarily so."

Getting a good handle on what you've got out there is step one, both in a traditional and cloud environment. A key next step is extreme standardization on both; if, for example, you know all SQL servers look like so-and-so, and only come in a small, medium or large, it's easier to grasp and manage the environment.

The same goes for dependencies; you can reduce them with techniques like inversion of control, and use of versioned API and data contracts.

Your enemy is irrational variation and dependencies, so rooting that out should be job one.

Think about how you spend your time each day. Why do you need to meet with others to discuss projects, or changes, or risks, etc.? There are two main drivers: variation, and dependencies. So root those out relentlessly to reduce the overhead of managing your environment.

IT-Led service configuration – practices

IT-Led Services
Configuration
Software
• Applications
• Data
Platform
• Runtime
• Middleware
• Operating System
Infrastructure
• Virtualization
• Server
• Storage
• Network
• Hardware
• Facilities

Dependency reduction

    <u>Inversion of control</u>

    <u>Versioned APIs</u>

    <u>Data contracts</u>

    <u>Microservices architecture / REST</u>

Variation reduction

    <u>Immutable deployment</u>

    <u>Infrastructure-as-code</u>

<u>Service mapping</u>

<u>Containerization</u>

Best practices for IT-Led service configuration are those that answer the question, "how do we know we have the right parts for our services, and that those parts are right, that is, as they should be, fitting to support the services they make up?"

Getting a good handle on what you've got out there is step one, both in a traditional and cloud environment. A key next step is extreme standardization on both; if, for example, you know all SQL servers look like so-and-so, and only come in a small, medium or large, it's easier to grasp and manage the environment.

The same goes for dependencies; you can reduce them with techniques like inversion of control, and use of versioned API and data contracts.

Your enemy is irrational variation and dependencies, so rooting that out should be job one.

Think about how you spend your time each day. Why do you need to meet with others to discuss projects, or changes, or risks, etc.? There are two main drivers: variation, and dependencies. So root those out relentlessly to reduce the overhead of managing your environment.

IT-Led service configuration – Software – definition

IT-Led Services

Configuration

Software
- Applications
- Data

Platform
- Runtime
- Middleware
- Operating System

Infrastructure
- Virtualization
- Server
- Storage
- Network
- Hardware
- Facilities

Application software (mode 1)

A program or group of programs designed for end users; application software resides above (operating) system software and includes applications such as database programs, word processors and spreadsheets.

Software as a Service (mode 2)

SaaS is a software delivery method that provides access to software and its functions remotely as a web-based service on a subscription basis. Contrast this with traditional application software, which is licensed to run locally on a device.

https://www.techopedia.com/definition/4224/application-software
04-01-05. service configuration – Software – Definition, desired state, best practices

Whether software is delivered in a traditional manner, or as a service, it must be the right software, fitting for the service(s) it supports.

IT-Led service configuration – Software – desired state

IT-Led Services
Configuration
Software ◄
• Applications
• Data
Platform
• Runtime
• Middleware
• Operating System
Infrastructure
• Virtualization
• Server
• Storage
• Network
• Hardware
• Facilities

Performance is effective when...

- We know what software we have, both licensed application software (mode 1) and SaaS (mode 2) and it is being used (we don't have any application software or SaaS we don't know about or aren't using)
- Software that is part of a service – we control its configuration in a known good state, and its functionality and service qualities meet what is required for the services by customers and users
- Software includes not just a UI, but an API and command line interface, so that it can be integrated
- Software is instrumented for direction and control by the SMS

04-01-05. service configuration – Software – Definition, desired state, best practices

Knowing what you have (software, licenses, etc.) is step one in understanding and managing your software configuration.

IT-Led service configuration - Software - practices

IT-Led Services
Configuration
Software
• Applications
• Data
Platform
• Runtime
• Middleware
• Operating System
Infrastructure
• Virtualization
• Server
• Storage
• Network
• Hardware
• Facilities

Best practices for software configuration include:
- ■ License management
- ■ SaaS software best practices
- ■ Cloud cost containment

An example: cloud cost containment.

It is typical now to find in organizations lots of unused seats, e.g., for Office365, that sit unused, or environments that were spun up in, e.g., AWS that nobody turned down. The problem is that the meter is running on this unused capability, and the resources going to waste there are typically sorely needed for things that actually will add value.

As a result, a new role is emerging in organizations for individuals and teams and technologies and policies that help identify dis-used resources and turn them off, and that make sure the organization has visualization into what is getting stood up and down, and used and not used. A further aspect of this role is identifying, from the services and feature sets and features being firehosed at the organization by vendors, what small subset of the are of high value, and therefore should be socialized within the organization to ensure value uptake. Similarly, for organizations that produce services, features, and feature sets and spew them at their own consumers, there is a need for a function in the provider organization to ensure uptake by consumers, to ensure value is as fully realized as possible.

IT-Led Services

Configuration

Software
• Applications
• Data

Platform
• Runtime
• Middleware
• Operating
  System

Infrastructure
• Virtualization
• Server
• Storage
• Network
• Hardware
• Facilities

Applications

Programs designed for end users. The licensing and delivery model for applications can be Mode 1 / purchased applications that run locally on the client device, or Mode 2 / Software as a Service (SaaS applications) that are cloud- and subscription-based where a supplier hosts applications and makes them available to customers over the Internet, e.g., Microsoft Office (Mode 1) / Office365 (Mode 2)

04-01-06. service configuration - Applications - Definition, desired state, best practices

Module 4: Service concepts, desired states and practices

89

Typically, at the time of this writing, most organization have a mix of traditional on-prem and locally running apps on devices and hosted or subscription-based apps.

IT-Led service configuration – Applications – desired states

IT-Led Services
Configuration
Software
• Applications
• Data
Platform
• Runtime
• Middleware
• Operating System
Infrastructure
• Virtualization
• Server
• Storage
• Network
• Hardware
• Facilities

Performance is effective when...

■ We have the right mix of Mode 1 / purchased applications that run locally on the client device, and Mode 2 / Software as a Service (SaaS applications) that are cloud- and subscription-based where a supplier hosts applications and makes them available to customers over the Internet

■ For applications we buy or subscribe to, costs are in control, and we do not have an inordinate amount of purchased or subscribed applications that go unused

■ For applications we buy or subscribe to, we know without undue effort, what new features sets and features are high value, and make sure they are taken up in the organization; for applications we make, we enable our consumers to do the same

04-01-06. service configuration – Applications – Definition, desired state, best practices

Uptake for realization of value is a top concern for both providers of apps and their consumers. Regardless of what functionality an app has, no value is realized if the functionality goes unused. As the pace of feature introduction increases dramatically with Agile and DevOps approaches, ensuring uptake because a key need and skill in provider and consumer organizations.

IT-Led service configuration – Applications – practices

IT-Led Services
Configuration

Software
• Applications
• Data

Platform
• Runtime
• Middleware
• Operating
  System

Infrastructure
• Virtualization
• Server
• Storage
• Network
• Hardware
• Facilities

- ■ Application Lifecycle Management (ALM)
- ■ Microservices architecture
- ■ Software configuration management

IT-Led service configuration – Data – definition

IT-Led Services
Configuration

Software
- Applications
- Data ◀

Platform
- Runtime
- Middleware
- Operating System

Infrastructure
- Virtualization
- Server
- Storage
- Network
- Hardware
- Facilities

Data

Data used by the application component of the service. The licensing and delivery model for applications can be Mode 1 / data for purchased applications that run locally on the client device, or data for Mode 2 / Software as a Service (SaaS applications) that are cloud- and subscription-based where a supplier hosts applications and makes them available to customers over the Internet, e.g., Microsoft Office (Mode 1) / Office365 (Mode 2)

IT-Led service configuration – Data – desired state

IT-Led Services

Configuration

Software
• Applications
• Data

Platform
• Runtime
• Middleware
• Operating
  System

Infrastructure
• Virtualization
• Server
• Storage
• Network
• Hardware
• Facilities

Performance is effective when...

■ Automation

04-01-07. service configuration – Data – Definition, desired state, best practices

Module 4: Service concepts, desired states and practices

IT-Led Services

Configuration

Software
• Applications
• Data

Platform
• Runtime
• Middleware
• Operating
  System

Infrastructure
• Virtualization
• Server
• Storage
• Network
• Hardware
• Facilities

- SaaS data integration
- Tenancy patterns
- Sharding
- Microservices
- REST

04-01-07. service configuration – Data – Definition, desired state, best practices

Module 4: Service concepts, desired states and practices

94

An example: Sharding.

Sharding is a type of database partitioning that separates very large databases the into smaller, faster, more easily managed parts called data shards, where a shard means a smaller part of a whole.

IT-Led service configuration – Platform – definition

IT-Led Services

Configuration

Software
• Applications
• Data

Platform ◀
• Runtime
• Middleware
• Operating
  System

Infrastructure
• Virtualization
• Server
• Storage
• Network
• Hardware
• Facilities

Platform

Provides the ability, in combination with infrastructure, to develop, test, run and host applications. A platform for computing typically includes operating system, programming language execution environment, database, web server, etc.

In Mode 2, PaaS (Platform as a Service), the platforms are cloud- and subscription-based where a supplier hosts the platform and makes it available to customers over the Internet, e.g., AWS Elastic Beanstalk, Windows Azure, Heroku, Force.com, Google App Engine, Apache Stratos.

Source: https://stackoverflow.com/questions/16820336/what-is-saas-paas-and-iaas-with-examples
04-01-08. service configuration – Platform – Definition, desired state, best practices

The platform is the runtime environment for software.

IT-Led service configuration – Platform – desired state

IT-Led Services

Configuration

Software
• Applications
• Data

Platform ◄
• Runtime
• Middleware
• Operating System

Infrastructure
• Virtualization
• Server
• Storage
• Network
• Hardware
• Facilities

Performance is effective when...

The platform compares well across the following criteria categories:

1. Characteristics - characteristics (i.e. on-demand self-service, resource pooling, rapid elasticity, and measured service)

2. Dimensions - how widely the solution can be shared (i.e. private, public, community), who is responsible for environment management (i.e. internal, external), and where the platform is located (i.e. on-prem, outsourced)

3. Production ready - platform suitability for enterprise, mission critical use

4. Development activities and lifecycle phases - Measures how to design, build, deploy, and manage applications and services (e.g., with DevOps practices: continuous integration / delivery, automated release, incremental testing)

5. Architecture - principles, concepts, and patterns enable applications to dynamically execute parallel workloads across a distributed environment

6. Platform services - how fully the platform satisfies development of complex applications with comprehensive application middleware components, services

7. Programming model - programming languages and frameworks, which facilitates building applications and services exhibiting the right characteristics

Source: http://wso2.com/wso2_resources/wso2-whitepaper-selecting-a-cloud-platform.pdf
04-01-08. service configuration – Platform – Definition, desired state, best practices

The typical environment at the time of this writing is a hybrid of traditional IT and cloud, so a mix of both kinds of platforms are in evidence.

IT-Led service configuration – Platform – practices

IT-Led Services
Configuration

Software
• Applications
• Data

Platform
• Runtime
• Middleware
• Operating System

Infrastructure
• Virtualization
• Server
• Storage
• Network
• Hardware
• Facilities

■ PaaS
■ Cloud
■ PaaS security

An example: PaaS security. The PaaS security link above cites what a PaaS service should be able to do:

• Scan and analyze mobile applications.

• Scan web apps on the internet or private networks.

• Perform static analysis to scan source code for security vulnerabilities.

• Employ single sign-on capabilities.

• Drain logs over the syslog, syslog-tls or HTTPS, including all the events related to the application.

• Distinguish logs from different instances of the same application.

• Encrypt data at rest.

• Facilitate secure communication between the application and database instance.

• Discover sensitive data and stored procedures for masking sensitive data.

IT-Led service configuration - Runtime - definition

IT-Led Services

Configuration

Software
• Applications
• Data

Platform
• Runtime ◀
• Middleware
• Operating
  System

Infrastructure
• Virtualization
• Server
• Storage
• Network
• Hardware
• Facilities

Runtime

The environment (software) that certain applications depend on to run on a device, that must be running for the application to execute. It provides common routines and functions that the applications require, and it typically converts the program, which is in an interim, intermediate language, into machine language. The runtime environment within which the application component of a service executes, e.g., Java RTE or JRE, Visual Basic runtime module, .NET Common Language Runtime (CLR) engine, Adobe's AIR runtime engine, Python runtime environment.

https://techterms.com/definition/rte
https://www.pcmag.com/encyclopedia/term/56079/runtime-engine

Software like Adobe Flash Player provides a runtime environment for its associated file format, allowing, in this case, Flash movies to be run within the player software.

IT-Led service configuration - Runtime - desired state

IT-Led Services

Configuration

Software
- Applications
- Data

Platform
- Runtime ◀
- Middleware
- Operating System

Infrastructure
- Virtualization
- Server
- Storage
- Network
- Hardware
- Facilities

Performance is effective when...

- The runtime environment provides an effective means for testing, debugging, and running the application component of a service, including provisions like crash dumps, step execution, and logging and monitoring.

Runtime refers to the runtime environment within which the application component of a service executes, e.g., Java RTE.

IT-Led service configuration – Runtime – practices

IT-Led Services
Configuration

Software
• Applications
• Data

Platform
• Runtime ◀
• Middleware
• Operating
  System

Infrastructure
• Virtualization
• Server
• Storage
• Network
• Hardware
• Facilities

■ <u>Runtime environment</u>

This example, while specific to a specific RTE, provide a good checklist that can be adapted for other RTEs.

IT-Led service configuration – Middleware – definition

IT-Led Services

Configuration

Software
- Applications
- Data

Platform
- Runtime
- Middleware
- Operating System

Infrastructure
- Virtualization
- Server
- Storage
- Network
- Hardware
- Facilities

Middleware

Software that serves to "glue together" separate, often complex and already existing, programs. Some software components that are frequently connected with middleware include enterprise applications and Web services, e.g., include database, application server, message-oriented, and web middleware, and transaction-processing monitors.

http://searchmicroservices.techtarget.com/definition/middleware

https://azure.microsoft.com/en-us/overview/what-is-middleware/

04-01-10. service configuration – Middleware – Definition, desired state, best practices

Middleware is the basis for integrating separate applications. For a good list of types of middleware, see https://apprenda.com/library/architecture/types-of-middleware.

IT-Led service configuration – Middleware – desired state

IT-Led Services
Configuration

Software
• Applications
• Data

Platform
• Runtime
• Middleware ◄
• Operating
  System

Infrastructure
• Virtualization
• Server
• Storage
• Network
• Hardware
• Facilities

Performance is effective when...

■ We don't have Developers building our production middleware configuration

■ We have put in place in-depth historical middleware and monitoring

■ We have scripted and versioned installation, configuration and deployment

Source: blogs.oracle.com/emeapartnerweblogic/5-best-practices-for-middleware-operations-teams-by-c2b2
04-01-10. service configuration – Middleware – Definition, desired state, best practices

Middleware is the software that serves to "glue together" separate, often complex and already existing, programs. Some software components that are frequently connected with middleware include enterprise applications and web services.

IT-Led service configuration – Middleware – practices

IT-Led Services

Configuration

Software
- Applications
- Data

Platform
- Runtime
- Middleware
- Operating System

Infrastructure
- Virtualization
- Server
- Storage
- Network
- Hardware
- Facilities

- **Middleware best practices**
- **Middleware checklist**

Middleware best practices cover provisioning, monitoring, configuration management, compliance, lifecycle management, and information publishing.

IT-Led Services
Configuration

Software
• Applications
• Data

Platform
• Runtime
• Middleware
• Operating System

Infrastructure
• Virtualization
• Server
• Storage
• Network
• Hardware
• Facilities

Operating System

The system software that runs on a computer (real or virtual) used to run other programs and applications. Computer operating systems perform basic tasks, such as recognizing input from the keyboard, sending output to the display screen, keeping track of files and directories on the disk, and controlling peripheral devices such as printers, making sure that different programs and users running at the same time do not interfere with each other, and ensuring that unauthorized users do not access the system.

http://www.webopedia.com/TERM/O/operating_system.html
04-01-11. service configuration – Operating system – Definition, desired state, best practices

Module 4: Service concepts, desired states and practices

104

Common operating systems include Windows (in a variety of versions), as well as iOS, Android, Macintosh, Chrome OS.

Containerization is OS virtualization in which the kernel allows the existence of multiple isolated user-space instances, called containers.

IT-Led service configuration – Operating system – desired state

IT-Led Services

Configuration

Software
• Applications
• Data

Platform
• Runtime
• Middleware
• Operating
  System

Infrastructure
• Virtualization
• Server
• Storage
• Network
• Hardware
• Facilities

Performance is effective when...

■ We have practices that work well for deploying and securing the OS, cleaning up unnecessary software, keeping OS instances up-to-date with service packs and patches, group policies, security templates, and configuration baselines.

Source: www.continuum.net/blog/6-important-steps-to-harden-your-clients-operating-systemshttps://en.wikipedia.org/wiki/List_of_system_quality_attributes

04-01-11. service configuration – Operating system – Definition, desired state, best practices

As you can see, these are basic hygiene practices that, if ignored, can lead to unwarranted overhead, at a minimum, or worse, a major incident or a disaster, because, e.g., you are not keeping your attack surface to a minimum.

IT-Led service configuration – Operating system – practices

IT-Led Services
Configuration
Software
• Applications
• Data
Platform
• Runtime
• Middleware
• Operating System ◄
Infrastructure
• Virtualization
• Server
• Storage
• Network
• Hardware
• Facilities

OS hardening
- Programs clean-up
- Use of service packs
- Patches and patch management
- Group policies
- Security templates
- Configuration baselines

Containerization

https://www.continuum.net/blog/6-important-steps-to-harden-your-clients-operating-systems

04-01-11. service configuration – Operating system – Definition, desired state, best practices

Two best practices for operating systems are OS hardening and containerization.

IT-Led service configuration – Infrastructure – definition

IT-Led Services
Configuration

Software
• Applications
• Data

Platform
• Runtime
• Middleware
• Operating System

Infrastructure
• Virtualization
• Server
• Storage
• Network
• Hardware
• Facilities

Infrastructure

Compute resources, along with storage and network capabilities. In Mode 2, it is Infrastructure as a service (IaaS), a standardized, highly automated offering, where compute resources, complemented by storage and networking capabilities are owned and hosted by a service provider and offered to customers on-demand. Customers are able to self-provision this infrastructure, using a Web-based graphical user interface that serves as an IT operations management console for the overall environment. API access to the infrastructure may also be offered as an option. IaaS provides compute infrastructure, physical or (often) virtual machines and other resources like virtual-machine disk image library, block and file-based storage, firewalls, load balancers, IP addresses, virtual local area networks etc., e.g., Amazon EC2, Windows Azure, Rackspace, Google Compute Engine.

Adapted from: http://www.gartner.com/it-glossary/infrastructure-as-a-service-iaas

https://stackoverflow.com/questions/16820336/what-is-saas-paas-and-iaas-with-examples

Your infrastructure underpins your platform and the software you run, whether that infrastructure is traditional / on-premise or in the cloud.

IT-Led service configuration – Infrastructure – desired state

IT-Led Services
Configuration
Software
• Applications
• Data
Platform
• Runtime
• Middleware
• Operating System
Infrastructure
• Virtualization
• Server
• Storage
• Network
• Hardware
• Facilities

Performance is effective when...

- We have the level of access we need to our infrastructure
- We know how much modification our software requires to run on it
- We know who owns the infrastructure we are paying for
- We are crystal clear on our (if we host) or the vendor's monitoring and support process
- We know our (if we host) or our vendor's backup plan and have made sure it is in the SLA process
- We understand the cost (if we host) or vendor's pricing model and how to control usage
- We have made sure we (if we host) or the vendor can ensure compliance with regulations

Source: http://www.zdnet.com/article/iaas-checklist-best-practices-for-picking-an-iaas-vendor/
04-01-12. service configuration – Infrastructure – Definition, desired state, best practices

Whether you in-source or outsource your infrastructure, and whether it is traditional or cloud infrastructure, you still need to make sure you have visibility into what it is, that it is monitored, backed up, what SLAs apply to it, what it costs, and whether or not it is in compliance with relevant regulations.

IT-Led service configuration - Infrastructure - practices

IT-Led Services
Configuration

Software
• Applications
• Data

Platform
• Runtime
• Middleware
• Operating
  System

Infrastructure
• Virtualization
• Server
• Storage
• Network
• Hardware
• Facilities

- IaaS
- Infrastructure-as-cloud
- IaaS security

An example: IaaS security. The source reference above indicates The security controls in an effective IaaS program should include the ability to:

- Manage data center identities and access.

- Authenticate, authorize and manage users.

- Secure and isolate virtual machines (VM).

- Patch default images for compliance.

- Monitor logs on all resources.

- Isolate networks.

IT-Led service configuration – Virtualization – definition

IT-Led Services

Configuration

Software
• Applications
• Data

Platform
• Runtime
• Middleware
• Operating
  System

Infrastructure
• Virtualization ◄
• Server
• Storage
• Network
• Hardware
• Facilities

Virtualization

The abstraction of IT resources that masks the physical nature and boundaries of those resources from resource users. An IT resource can be a server, a client, storage, networks, applications or OSs. Essentially, any IT building block can potentially be abstracted from resource users.

Source: http://www.gartner.com/it-glossary/virtualization/
04-01-13. service configuration – Virtualization – Definition, desired state, best practices

Top virtualization platforms as of this writing include VMware vSphere, Microsoft Hyper-V, Citrix XenServer, and Oracle VM.

IT-Led service configuration – Virtualization – desired state

IT-Led Services
Configuration

Software
• Applications
• Data

Platform
• Runtime
• Middleware
• Operating System

Infrastructure
• Virtualization
• Server
• Storage
• Network
• Hardware
• Facilities

Performance is effective when...

■ We know the advantages and disadvantages of virtualization
■ We know the performance bottlenecks of each system role
■ We don't over-prioritize management, patching and security of virtual systems
■ We don't treat virtual systems differently than physical systems unless absolutely necessary
■ We backup early, backup often
■ We are careful when using any "undo" technology
■ We understand our failover and our scale-up strategy
■ We control virtual machine proliferation
■ We centralize our storage
■ We understand our security perimeter

Source: https://technet.microsoft.com/en-us/library/gg131921.aspx
04-01-13. service configuration – Virtualization – Definition, desired state, best practices

One key thing here is managing VM proliferation—just like physical server sprawl, it is a must to keep on top of what has been spun up and is not dis-used.

IT-Led service configuration – Virtualization – practices

IT-Led Services
Configuration

Software
• Applications
• Data

Platform
• Runtime
• Middleware
• Operating
  System

Infrastructure
• Virtualization
• Server
• Storage
• Network
• Hardware
• Facilities

- ■ <u>Virtualization / Hypervisor</u>
  - ■ Tuning
  - ■ Patching
  - ■ Security
  - ■ Backup
  - ■ Failover
  - ■ Scale-up
  - ■ Controlling proliferation
  - ■ Storage centralization
  - ■ Securing the perimeter
- ■ <u>Container management and orchestration</u>

Source: adapted from http://searchitoperations.techtarget.com/definition/container-management-software
04-01-13. service configuration – Virtualization – Definition, desired state, best practices
Module 4: Service concepts, desired states and practices

Container management is handling tasks associated with the administration of individual containerized applications and application components deployed on individual hosts. Applications are packaged into containers for ease of scaling, duplication and upgrading.

Container orchestration is managing multiple containers deployed on multiple hosts, extending lifecycle management capabilities to complex, multi-container workloads deployed on a cluster of machines.

IT-Led service configuration – Server – definition

IT-Led Services
Configuration

Software
• Applications
• Data

Platform
• Runtime
• Middleware
• Operating
  System

Infrastructure
• Virtualization
• Server ◄
• Storage
• Network
• Hardware
• Facilities

Server

A computer, a device or a program that is dedicated to managing network resources. Servers are often referred to as dedicated because they carry out hardly any other tasks apart from their assigned roles. There are a number of categories of servers, including print servers, file servers, network servers and database servers. In theory, whenever computers share resources with client machines they are considered servers. Servers can be physical or virtual.

Source: Adapted from https://www.techopedia.com/definition/2282/server
04-01-14. service configuration – Server – Definition, desired state, best practices

Module 4: Service concepts, desired states and practices

For an example of the different kinds of roles a server can take on, see https://technet.microsoft.com/en-us/library/hh831669(v=ws.11).aspx.

IT-Led service configuration - Server - desired state

IT-Led Services
Configuration

Software
• Applications
• Data

Platform
• Runtime
• Middleware
• Operating System

Infrastructure
• Virtualization
• Server ◀
• Storage
• Network
• Hardware
• Facilities

Performance is effective when...

■ We know of (and can quickly and easily visualize) each server that we have, and the configuration of each server, to a level of detail that is useful to us

■ We maintain strict control over server build templates

■ We relentlessly root out unnecessary server complexity, variation and dependencies

04-01-14. service configuration - Server - Definition, desired state, best practices

In general, we want to avoid the situation of "snowflake" servers—where servers are unique—to avoid the overhead associated with variation—in term of the time and effort required to sort out the risks of variations and dependencies when we make changes, and the cost associated with mistakes due to missed variation and dependencies.

IT-Led service configuration – Server – practices

IT-Led Services

Configuration

Software
• Applications
• Data

Platform
• Runtime
• Middleware
• Operating System

Infrastructure
• Virtualization
• Server
• Storage
• Network
• Hardware
• Facilities

■ <u>Server Management</u>

#10. Implement a Regular Maintenance Schedule
#9. Automate Everything + Manage by Exception
#8. Run Weekly Windows Updates + Install All Security Patches
#7. REBOOT
#6. Housekeeping
#5. Diskspace, Defrag and Memory
#4. Inventory of Running Software
#3. Stagger Updates
#2. Run During Weekdays
#1. Report Results

■ <u>Visible Ops-style change control</u>

■ <u>Continuous delivery</u>

04-01-14. service configuration - Server - Definition, desired state, best practices

For example, #9. Automate Everything + Manage by Exception: Almost any task that you run at the server console can be automated and scheduled. If you do any maintenance regularly, or at least four times per year, automate it, for two reasons:

1. Once automated, it is much less prone to human errors. Human errors or oversights are a big reason why maintenance is not performed, or performed incorrectly.

2. Automated tasks have much better Run Tracking Logs for history and troubleshooting.

Once these maintenance tasks are automated, you can Manage by Exception and only work on failed server upgrades when, for example, a server does not return to service after a reboot.

What you want with servers is what DevOps practitioners refer to as cattle, not pets. The Visible Ops Handbook, and the book by Jez Humble, Continuous Delivery, both discuss the dangers of snowflakes and how to avoid them, the first in the context of a solid basis for effective change control, the latter as a solid basis for an effective build and delivery process.

IT-Led service configuration – Storage – definition

IT-Led Services

Configuration

Software
• Applications
• Data

Platform
• Runtime
• Middleware
• Operating System

Infrastructure
• Virtualization
• Server
• Storage ◀
• Network
• Hardware
• Facilities

Storage

Repository for digital data is saved within a data storage device by means of computing technology; mechanism that enables a computer to retain data, either temporarily or permanently; may also be referred to as computer data storage or electronic data storage.

Source: Adapted from https://www.techopedia.com/definition/1115/storage
04-01-15. service configuration – Storage – Definition, desired state, best practices

Storage needs to be managing, just as compute and network, and regardless of if that storage is traditional on-premise storage or in the cloud.

IT-Led service configuration – Storage – desired state

IT-Led Services

Configuration

Software
- Applications
- Data

Platform
- Runtime
- Middleware
- Operating System

Infrastructure
- Virtualization
- Server
- Storage ◀
- Network
- Hardware
- Facilities

Performance is effective when...

- We leverage tiered storage
- We analyze application workloads and make educated decisions on where to store data
- We consolidate storage pools
- We implement staged backup to disk
- We employ automated storage management tools

https://esj.com/articles/2009/10/27/best-storage-practices.aspx
04-01-15. service configuration – Storage – Definition, desired state, best practices

IT-Led Services

Configuration

Software
• Applications
• Data

Platform
• Runtime
• Middleware
• Operating System

Infrastructure
• Virtualization
• Server
• Storage
• Network
• Hardware
• Facilities

Storage best practices
- Tiered storage
- Workload analysis
- Storage pooling
- Staged backup
- Automated storage management

Cloud storage best practices

According the link above, cloud storage considerations include:

- Getting the degree of redundancy right (multi-site redundancy, single-site redundancy (mirroring, etc.) or none)

- Automatic fail-over in case of a disk/server/site failure

- Versioning capability, not just storage of the most current version of a file or data object, with the right retention period for deleted files, and a flexible retention policy

- Data backup, with the right backup cycle and retention policy, and the right time-to-restore data

- An easy-to-use management console that is web-based and can be accessed from any location in case of emergency

- A pricing structure that fits your business model? For example, some vendors charge for every file access (read, write, etc.) in addition to per-gigabyte upload and download charges. If you are moving large blocks of data those access charges will be minimal. If you are doing primarily database lookups and updates, however, they can add up quickly.

IT-Led service configuration – Network – definition

IT-Led Services
Configuration

Software
• Applications
• Data

Platform
• Runtime
• Middleware
• Operating
  System

Infrastructure
• Virtualization
• Server
• Storage
• Network ◀
• Hardware
• Facilities

Network

A group of two or more devices that can communicate. In practice, a network is comprised of a number of different computer systems connected by physical and/or wireless connections. The scale can range from a single device sharing out basic peripherals to large data centers located worldwide, to the Internet itself. Regardless of scope, all networks allow devices and people to share information and resources.

Source: Adapted from https://www.techopedia.com/definition/5537/network
04-01-16. service configuration – Network – Definition, desired state, best practices

Physical and virtual and cloud networking resources must be known and managed to ensure the services they depend on are as they should be.

IT-Led service configuration – Network – desired state

IT-Led Services

IT-Led Services

Configuration

Software
• Applications
• Data

Platform
• Runtime
• Middleware
• Operating
  System

Infrastructure
• Virtualization
• Server
• Storage
• Network ◀
• Hardware
• Facilities

Performance is effective when...

■ Our network is made up of components that support the level of, e.g., availability, performance, and security our services require

04-01-16. service configuration – Network – Definition, desired state, best practices

Module 4: Service concepts, desired states and practices

120

Networks components must meet the levels of availability, performance, and security services require. This is not possible if the components they are comprised of do not. For example, you will never get 5 nines on a 1 nine network, or a network that lacks redundancies in switches, power supplies, and so on.

IT-Led service configuration – Network – practices

IT-Led Services

Configuration

Software
• Applications
• Data

Platform
• Runtime
• Middleware
• Operating
  System

Infrastructure
• Virtualization
• Server
• Storage
• Network ◄
• Hardware
• Facilities

Network configuration management
- Administrator Approval for Changes
- Auto Discovery
- Automated Software Distribution
- Automatic Network Mapping
- Backup and Restore
- Baseline NRA
- Bulk Configuration Changes
- Change Management
- Compliance Audits/Reports
- Configuration Archive
- Configuration Compare
- Configuration Templates
- Copy Configuration

- Inventory/Asset Management
- Multi-vendor Device Support
- Network Provisioning
- Pre-Provisioning
- Real-time Change Notifications
- Remote Configuration
- Resource Initialization
- Resource Shutdown/Startup
- Scheduled Backups
- Scheduled Configuration Changes
- Scheduled Device
  Shutdown/Startup
- Scheduled Tasks

04-01-16. service configuration – Network - Definition, desired state, best practices

Module 4: Service concepts, desired states and practices

121

An example: reducing variation in networking equipment is key. For example, if you have several models of routers or switches, all with different defaults, it will be easy to make a mistake when configuring a device initially. Better to make it all "vanilla", or whatever flavor you like, to avoid the time it take to sort out differences and the cost of recovering from mistakes made because of variations.

IT-Led service configuration – Hardware – definition

IT-Led Services
Configuration

Software
• Applications
• Data

Platform
• Runtime
• Middleware
• Operating
  System

Infrastructure
• Virtualization
• Server
• Storage
• Network
• Hardware ◀
• Facilities

Hardware

The physical elements that make up a computer or electronic system and everything else involved that is physically tangible. This includes the monitor, hard drive, memory and the CPU. Hardware works hand-in-hand with firmware and software to make a computer function.

Source: https://www.techopedia.com/definition/2210/hardware-hw
04-01-17. service configuration – Hardware – Definition, desired state, best practices

Hardware is not of concern for IaaS and the cloud, at least not for the consumer. But for traditional environments, managing physical hardware is part of the knitting of what things get done.

IT-Led service configuration – Hardware – desired state

IT-Led Services

Configuration

Software
• Applications
• Data

Platform
• Runtime
• Middleware
• Operating
  System

Infrastructure
• Virtualization
• Server
• Storage
• Network
• Hardware
• Facilities

Performance is effective when...

■ We know of all the hardware our services depend on, along with its configuration to a level of detail that is useful to us, and this information is visible and readily accessible

■ We can track what hardware belongs to which service(s), along with any other relationships and attributes that are useful, e.g., what P.O. introduced it, whether it is under warrantee, whether it is a leased or bought asset, etc.

■ We introduce new hardware and retire hardware at a pace that matches business needs; we proactively retire hardware before it becomes problematic for us

■ We perform physical maintenance (e.g., changing filters) on an appropriate schedule to avoid issues

04-01-17. service configuration – Hardware – Definition, desired state, best practices

Hardware assets require maintenance that virtual assets do not.

IT-Led Services
Configuration
Software
• Applications
• Data
Platform
• Runtime
• Middleware
• Operating System
Infrastructure
• Virtualization
• Server
• Storage
• Network
• Hardware
• Facilities

- Hardware IT asset management
- IT asset management and the cloud

The link above on ITAM and the cloud emphasizes a number of open question that highlight the disruption the cloud is causing in IT asset management processes born into and based on a physical, on-prem situation:

- Who is responsible? For deploying servers into the Cloud, monitoring costs in the Cloud, managing the "hardware" specification of servers in the Cloud?

- And, "How do we control?", who can deploy services in the Cloud, the level and size of services deployed in the Cloud, and what licenses are used in the Cloud?

IT-Led service configuration – Facilities – definition

IT-Led Services

IT-Led Services

Configuration

Software
• Applications
• Data

Platform
• Runtime
• Middleware
• Operating
  System

Infrastructure
• Virtualization
• Server
• Storage
• Network
• Hardware
• Facilities

Facilities

The physical service environment, including power conditioning / UPS, cooling, air flow, etc.

With the cloud and IaaS, facilities management becomes the concern of the host vendor, not the consumer. For any on-prem equipment, including wiring closets, data centers, and so on, this is the concern of the provider.

IT-Led service configuration – Facilities – desired state

IT-Led Services
Configuration
Software
• Applications
• Data
Platform
• Runtime
• Middleware
• Operating
  System
Infrastructure
• Virtualization
• Server
• Storage
• Network
• Hardware
• Facilities

Performance is effective when...

■ Our facilities are constructed of components that "add up" to supporting the right levels of availability, recoverability, performance, and so on, that our services require
■ Our physical equipment is compatible with our applications
■ We have a fast refresh schedule and detailed roadmap, and a coordinated equipment refresh process
■ We have experienced local staff and expert support
■ We have good management and performance tools

http://searchdatacenter.techtarget.com/tip/A-data-center-checklist-for-facility-design-and-IT-ops
04-01-18. service configuration – Facilities – Definition, desired state, best practices

Module 4: Service concepts, desired states and practices

Ignore the physical layer at your peril; hardware needs to operate within set tolerances, and therefore needs power conditioning to avoid over and under-voltage conditions, backup power for emergencies, the right air conditioning and air flow to meet equipment requirements for operating temperatures, and sensors and filters and the like to avoid exposing the equipment to things that would harm it, like airborne contaminants, water, and the like.

IT-Led Services
Configuration

Software
• Applications
• Data

Platform
• Runtime
• Middleware
• Operating
  System

Infrastructure
• Virtualization
• Server
• Storage
• Network
• Hardware
• Facilities

■ Configuration management
■ Facilities management

04-01-18. service configuration – Facilities – Definition, desired state, best practices

Module 4: Service concepts, desired states and practices

127

---

http://www.facilitiesnet.com/datacenters/article/8-Ways-To-Bring-Down-Data-Centers--17445 cites 8 ways to bring down datacenters:

1. A standby/emergency generator fails to start when utility power fails for more than a few seconds, or generator transfer switchgear fails to transfer, or the generator fails to run until utility power is restored.

2. A UPS system randomly fails even though utility power is good, and then fails to successfully transfer to bypass.

3. A UPS system or batteries fail when utility power fails, and the generator (if available) has not yet started and run up to speed, or during transfer between generator and utility sources.

4. Circuit breaker nuisance tripping.

5. Circuit breakers installed in 24/7 live switchgear that have not been recently cycled (opened and closed) or energized (design voltage applied) or tested can be problems waiting to surface at the worst time

6. Neutral and grounding issues.

7. Bypass and transfer mechanisms that have not recently been operated, or operated under load, or operated at all (even many years after installation).

8. Emergency power off (EPO) circuitry for 24/7 live facilities, which have not been recently tested or where validated (trusted) wiring diagrams are not available.

Lesson Summary

- The configuration of a service is one of three things worth managing, next to service functionality and qualities
- There are two types of services, Human-Led and IT-Led; each has within it a set of components for which we must aim to achieve and maintain a desired state through best practices
- Human-Led service configuration consists of people, service IP, service kits, collateral, systems & tools, goods and facilities
- IT-Led services configuration consist of software, platforms, and infrastructure; for mode 1, these can be on-prem and a combination of physical and virtual; for mode 2, these are, "...as a service"—IaaS, SaaS and PaaS

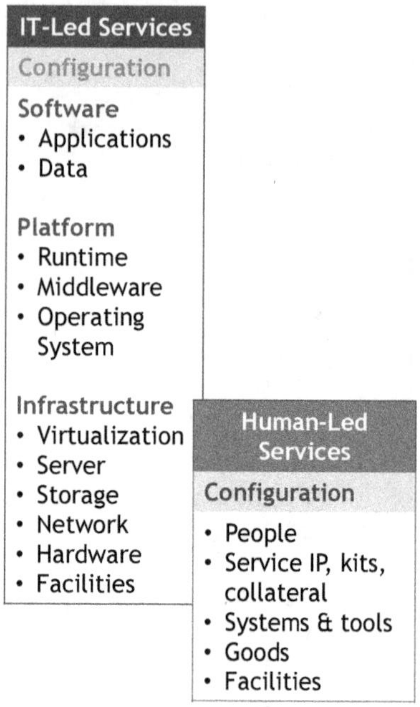

IT-Led Services

Configuration

Software
- Applications
- Data

Platform
- Runtime
- Middleware
- Operating System

Infrastructure
- Virtualization
- Server
- Storage
- Network
- Hardware
- Facilities

Human-Led Services

Configuration
- People
- Service IP, kits, collateral
- Systems & tools
- Goods
- Facilities

Figure 04-01-sum.1 IT-Led & Human Led services configuration 2

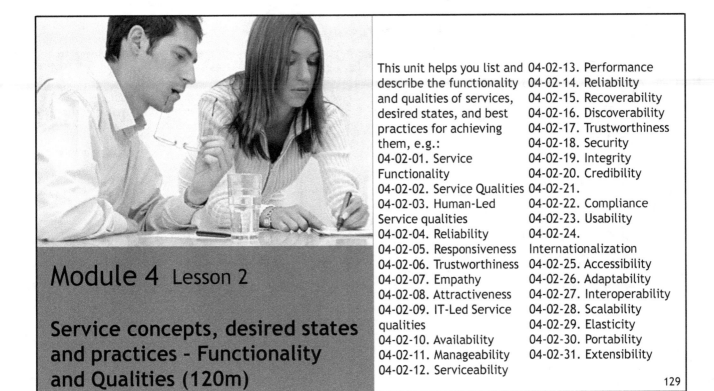

Module 4 Lesson 2

Service concepts, desired states and practices – Functionality and Qualities (120m)

This unit helps you list and describe the functionality and qualities of services, desired states, and best practices for achieving them, e.g.:

04-02-01. Service Functionality
04-02-02. Service Qualities
04-02-03. Human-Led Service qualities
04-02-04. Reliability
04-02-05. Responsiveness
04-02-06. Trustworthiness
04-02-07. Empathy
04-02-08. Attractiveness
04-02-09. IT-Led Service qualities
04-02-10. Availability
04-02-11. Manageability
04-02-12. Serviceability
04-02-13. Performance
04-02-14. Reliability
04-02-15. Recoverability
04-02-16. Discoverability
04-02-17. Trustworthiness
04-02-18. Security
04-02-19. Integrity
04-02-20. Credibility
04-02-21.
04-02-22. Compliance
04-02-23. Usability
04-02-24. Internationalization
04-02-25. Accessibility
04-02-26. Adaptability
04-02-27. Interoperability
04-02-28. Scalability
04-02-29. Elasticity
04-02-30. Portability
04-02-31. Extensibility

This is lesson two of module 4. We're focused here on the functionality (features) and qualities (what Devs call non-functional requirements, or the "-ilities"—availability, capacity and the like).

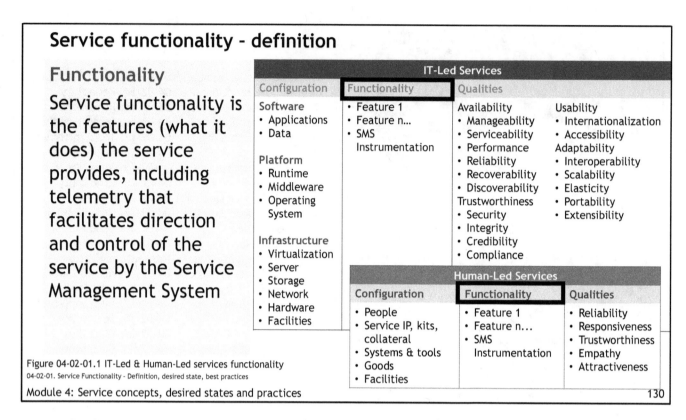

Service functionality - definition

Functionality

Service functionality is the features (what it does) the service provides, including telemetry that facilitates direction and control of the service by the Service Management System

IT-Led Services			
Configuration	Functionality	Qualities	
Software • Applications • Data Platform • Runtime • Middleware • Operating System Infrastructure • Virtualization • Server • Storage • Network • Hardware • Facilities	• Feature 1 • Feature n... • SMS Instrumentation	Availability • Manageability • Serviceability • Performance • Reliability • Recoverability • Discoverability Trustworthiness • Security • Integrity • Credibility • Compliance	Usability • Internationalization • Accessibility Adaptability • Interoperability • Scalability • Elasticity • Portability • Extensibility

Human-Led Services		
Configuration	Functionality	Qualities
• People • Service IP, kits, collateral • Systems & tools • Goods • Facilities	• Feature 1 • Feature n... • SMS Instrumentation	• Reliability • Responsiveness • Trustworthiness • Empathy • Attractiveness

Figure 04-02-01.1 IT-Led & Human-Led services functionality
04-02-01. Service Functionality - Definition, desired state, best practices

Module 4: Service concepts, desired states and practices   130

A service is made up of its component parts, its functionality (or features), and its qualities.

For a Human-Led service, you can think of it this way: Let's say you have a landscaping service. The configuration of the service is all the moving parts—your van, mower, leaf blower, edger, string trimmer, your clipboard, the phone you use to take payment, and so on. The functionality is the feature set of your service—does it include edging? Do you do weeding, or fertilizing, and so on. The qualities are things like how reliable and responsive you are, as judged by your customers

For IT-Led services, components are things like containers, servers, active directory, etc. Let's say your service was a financial calculator website—a feature might be a removal calculator, a margin calculator, or a mortgage payment calculator. The service qualities would be things like how available your website is (uptime) and how performant it is (fast or slow).

Functionality

Functionality of a services is what the services does, how it functions, it's basic features, what it does, independent of its behavioral characteristics. There are two types of functionality: 1) user functionality and 2) SMS telemetry; the latter is the telemetry built into services so that they can be directed and controlled by the Service Management System (SMS). Contrast service functionality or features with non-functional service qualities, which are how a service behaves, or all service characteristics that are not functional requirements, and with configuration, which are the parts a service is made up of.

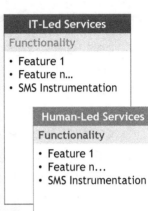

Figure 04-02-01.2 IT-Led & Human-Led services functionality 2
04-02-01. Service Functionality - Definition, desired state, best practices

Module 4: Service concepts, desired states and practices

131

So functionality is the set of features of a service, what it does, independent of how well it performs, how available it is, etc.

There are two types of functionality. First, there is what the user sees and uses, the user functionality. Next, there is the telemetry functionality—the features you bolt on to a service so that you can direct and control it, through whatever service management system it is that you have, for example, if you have a landscaping company, a number of crews out doing work on a given day, what do you have in place to tell if a particular crew has a client cancel on them and is available for work so you can re-direct them to another job? If you have an financial calculator website, what have you put in place to tell you if the site is performing slowly, or has an extraordinary number of concurrent users on it, and so on?

Remember, functionality is features, what a service does, and qualities are how a service behaves, like how available it is, how performant it is, and so on. You can also think of service qualities as all service characteristics that are not features or functionality.

Service functionality - desired state

Performance is effective when...

- Our services have the user features required by customers and users; where this is not the case, we move quickly and transparently to make it so, or to reset customer and user expectations and perceptions for the service functionality.

- Our services have the telemetry needed to direct and control them.

- Our services meet the minimally agreed to feature and viability requirements, operates within the principles, governance, and security boundaries of the organization, and are built and maintained by the revenue stream that needs the service, and are constantly evolving using automated testing and delivery techniques, with feature list and prioritization based on direct end-user feedback.

IT-Led Services

Functionality

- Feature 1
- Feature n...
- SMS Instrumentation

Human-Led Services

Functionality

- Feature 1
- Feature n...
- SMS Instrumentation

04-02-01. Service Functionality - Definition, desired state, best practices

Both end user functionality, and management functionality (telemetry) are required for a manageable service that is consistently valuable to stakeholders.

Service functionality - practices

- Agile requirements modeling
- Epics, stories, versions, and sprints

IT-Led Services
Functionality
• Feature 1
• Feature n...
• SMS Instrumentation

Human-Led Services
Functionality
• Feature 1
• Feature n...
• SMS Instrumentation

Whatever the method used, it's important that the outcome is that customers and users get functionality they value quickly, and with quality.

Lesson summary

- The functionality (features, what it does) of a service is one of three things worth managing, next to the service's configuration (component parts / what it is made of) and its qualities (how it behaves)

- There are two types of services, Human-Led and IT-Led; each has within it a set of features for which we must aim to achieve and maintain a desired state through best practices

- A critical part of a service's functionality is its telemetry embedded for direction and control by the service management system (SMS)

IT-Led Services

Functionality

- Feature 1
- Feature n...
- SMS Instrumentation

Human-Led Services

Functionality

- Feature 1
- Feature n...
- SMS Instrumentation

Service qualities – overall definition

Qualities

Service qualities are how a service behaves, its non-functional characteristics. Contrast this with the functionality of a service, which is what the services does, (what it does, how it functions, it's basic features independent of a service's behaviors), and the service configuration (what its made of, its component parts).

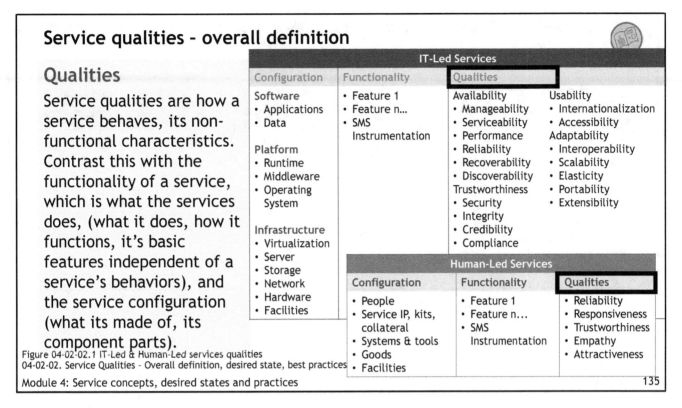

Figure 04-02-02.1 IT-Led & Human-Led services qualities
04-02-02. Service Qualities – Overall definition, desired state, best practices

In contrast to a service's functionality, or features, that is, what it does, service qualities are how a service behaves.

As you can see in the figure, both Human-Led and IT-Led services have system qualities, just as they have features.

The Human-Led service quality taxonomy in OSM are adapted from the SERVQUAL model, which is commonly applied to Human-Led services. The IT-Led service quality taxonomy is adapted from the Open Group's taxonomy of non-functional requirements.

Each of these will be defined in turn as we move through this module.

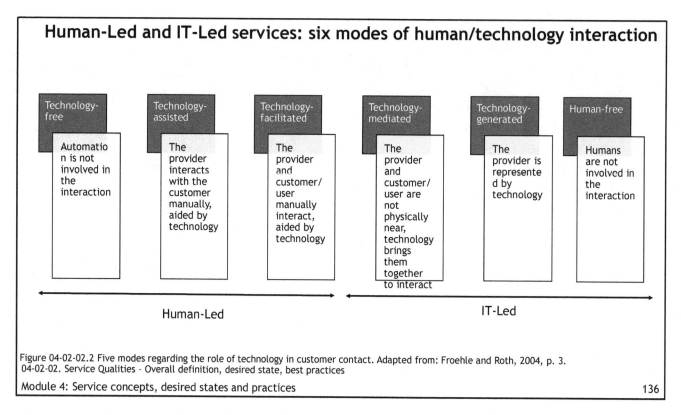

Figure 04-02-02.2 Five modes regarding the role of technology in customer contact. Adapted from: Froehle and Roth, 2004, p. 3.
04-02-02. Service Qualities – Overall definition, desired state, best practices

Module 4: Service concepts, desired states and practices   136

Froehle and Roth identify five modes of human and technology interaction.

1. Technology-free customer contact, e.g., the face-to-face contact between a psychological therapist and patient. This type of contact does not require the use of technology; thus, it is the most traditional and direct mode of customer contact.

2. Technology-assisted customer contact, e.g., hotel check-in and check-out procedures, transactions conducted over manned bank counters, and passenger check-ins for airline boarding. In these, technology (i.e., a computer) is used by the service personnel only. However, the customer and service personnel still experience face-to-face contact.

3. Technology-facilitated customer contact, e.g., the use of Microsoft PowerPoint by a financial expert to present and discuss financial plans with customers during a conference. Although technology is used by both parties, face-to-face customer contact still occurs.

4. Technology-mediated customer contact, e.g., telephonic interaction between service personnel and customers and the provision of professional advice by a consulting company using Videoconference technology. In these contexts, a shared technological platform is used by the service personnel and the customer without face-to-face contact.

5. Technology-generated customer contact, e.g., ATM withdrawals, online shopping. In these, customers operate the technology without the assistance of service personnel, and face-to-screen contact replaces face-to-face customer contact. We add a 6th one here, Human-free, where technology represents both the provider and the consumer of the service, as in EDI.

The point here is that services run on a continuum from technology-free to technology generated; Human-Led services begin at the left hand side of this continuum, ,up to the point where technology is the primary means of representing the service; IT-Led services start on the right, up to the point where humans are the primary means of representing the service.

Service qualities – overall desired state

Performance is effective when...

■ The service qualities required by customers and users are present in the service; where this is not the case, we move quickly and transparently to make it so, or to reset customer and user expectations and perceptions for the service qualities.

IT-Led Services	
Qualities	
Availability	Usability
• Manageability	• Internationalization
• Serviceability	• Accessibility
• Performance	Adaptability
• Reliability	• Interoperability
• Recoverability	• Scalability
• Discoverability	• Elasticity
Trustworthiness	• Portability
• Security	• Extensibility
• Integrity	
• Credibility	
• Compliance	

Human-Led Services
Qualities
• Reliability
• Responsiveness
• Trustworthiness
• Empathy
• Attractiveness

Figure 04-02-02.3 IT-Led & Human-Led services qualities 2
04-02-02. Service Qualities – Overall definition, desired state, best practices

Module 4: Service concepts, desired states and practices

Service qualities have a technical aspect and a qualitative aspect. For example, a service might, for accessibility, be ADA compliant, technically speaking. However, the customer and user are the judge of whether or not the system is accessible.

Service qualities – overall best practices

- Service Level Management
- Non-functional requirement framework
- Non-functional requirements specifications
- Non-function requirements checklist

IT-Led Services

Qualities

Availability
- Manageability
- Serviceability
- Performance
- Reliability
- Recoverability
- Discoverability

Trustworthiness
- Security
- Integrity
- Credibility
- Compliance

Usability
- Internationalization
- Accessibility

Adaptability
- Interoperability
- Scalability
- Elasticity
- Portability
- Extensibility

Human-Led Services

Qualities
- Reliability
- Responsiveness
- Trustworthiness
- Empathy
- Attractiveness

IT-Led service qualities adapted from: https://publications.opengroup.org/w098
04-02-02. Service Qualities – Overall definition, desired state, best practices

You should notice two major differences here between OSM and traditional ITSM guidance.

The first is that OSM covers Human-Led services, and their non-functional characteristics or qualities, where traditional ITSM guidance does not.

The second is that the list of non-functional requirements (what traditional ITSM guidance calls, "warranty" aspects) is much longer in OSM. Traditional ITSM guidance typically focused on operational qualities, things that went in an SLA, e.g., availability, performance (or capacity), IT service continuity (or disaster recovery), security, and supplier details.

As you can see, OSM's list of service qualities greatly expands the list to include a "shift-left" to the non-functional qualities developers care about, such as accessibility.

OSM's list of IT-Led service quality characteristics are adapted from the TOGAF taxonomy of service qualities.

Human-Led Services
Qualities
• Reliability
• Responsiveness
• Trustworthiness
• Empathy
• Attractiveness

Human-Led Service Qualities

Human-Led services are driven by humans, and may be assisted or facilitated by IT, for example, PC repair, Moves/Adds/Changes; contrast this with IT-Led services, which are driven by technology, and may be assisted or facilitated by humans, e.g., Office365. Service qualities are how a service behaves, its non-functional characteristics. Contrast this with the functionality of a service, which is what the services does, how it functions, it's basic features independent of a service's behaviors, and the service configuration (what its made of, its component parts). Human-Led service qualities include reliability, responsiveness, trustworthiness, empathy, and attractiveness.

Figure 04-02-03.1 Human-Led services qualities
04-02-03. Service Qualities – Human-Led Services - Definition, desired state, best practices

Module 4: Service concepts, desired states and practices

139

Human-Led services run the gamut from:

1. Technology-free customer contact, e.g., the face-to-face contact between a psychological therapist and patient. This type of contact does not require the use of technology; thus, it is the most traditional and direct mode of customer contact.

2. Technology-assisted customer contact, e.g., hotel check-in and check-out procedures, transactions conducted over manned bank counters, and passenger check-ins for airline boarding. During these, technology (i.e., a computer) is used by the service personnel only. However, the customer and service personnel still experience face-to-face contact.

3. Technology-facilitated customer contact, e.g., the use of Microsoft PowerPoint by a financial expert to present and discuss financial plans with customers during a conference. Although technology is used by both parties, face-to-face customer contact still occurs.

Human-Led Services

Qualities
- Reliability
- Responsiveness
- Trustworthiness
- Empathy
- Attractiveness

Human-Led Service Qualities

How a Human-Led service behaves, it's non-functional characteristics, including how reliable, responsive, trustworthy, empathetic, and attractive it is to customers and users.

Examples of Human-Led services
- Moves, adds and changes
- New hire setup
- Legal hold
- Consulting
- Auditing

Source: Human-Led service qualities in OSM are adapted from SERVQUAL
04-02-03. Service Qualities - Human-Led Services - Definition, desired state, best practices

Module 4: Service concepts, desired states and practices   140

Human-Led services are provided primarily by humans, possibly with the assistance of technology, but where the human-to-human interaction is the primary driver. Contrast this with IT-Led Services, where services are provided primary by technology, possibly with the assistance of humans, but where technology-to-technology or technology to human interaction is the primary driver. Human-Led service qualities in OSM are adapted from SERVQUAL.

Key questions for Human-Led service qualities include:

- People—the individuals and teams that deliver the service—how reliable and responsive are they? Are they dressed appropriately for what they're doing, from the perspective of customers and users?

- Service IP, kits, collateral—the knowledge and materials teams rely on to deliver services--for example, if you are running workshops as part of the service, are the materials attractive and well-organized, or do you have a lot of bad graphics and spelling errors in them?

- Systems & Tools—for example, process mapping software for a process engineering engagement—is the tool easy to use for the facilitator and the customer, or is it clunky and error-prone?

- Goods—for services where goods are involved, e.g., installing a new physical monitor—is your packaging for goods neat and complete, or do goods show up dented or broken or in the wrong place because of shoddy addressing?

- Facilities—is your service location neat, organized, appealing, or shoddy / off-putting?

Human-Led service quality – desired state

Human-Led Services

Qualities
- Reliability
- Responsiveness
- Trustworthiness
- Empathy
- Attractiveness

- Performance is effective when...
- Stakeholders see us and our services as reliable, responsive, trustworthy, empathetic, and attractive
- People—the individuals and teams that deliver the service, have the knowledge, skills and mindset required
- Service IP, kits, collateral—the knowledge and materials teams rely on to deliver services—is up to date and easy to location
- Systems & Tools—for example, a process template for a process engineering engagement—exist and are good enough to ensure successful delivery
- Goods—for services with goods involved, e.g., installing a new physical monitor—we manage inventory to meet demand
- Facilities—are physically appealing, and strongly support the conduct of the service

Source: https://en.wikipedia.org/wiki/SERVQUAL
04-02-03. Service Qualities – Human-Led Services - Definition, desired state, best practices

"Attractive" as used here does not mean that all our people are supermodels, that all our facilities are slick, etc. It means at a minimum that they are not off-putting, and are suitable for the purposes of the service and its stakeholders. For example, there is one standard of dress you might expect a dollar store employee to have, and quite another for say, a Nordstrom salesperson. And if the dollar store fixtures were too expensive-looking, you might wonder what you're getting for your dollar—similarly, if the fixtures in a Nordstrom's were shoddy or cheap looking, you might be wondering if the t-shirt you're buying is really worth $30.

Human-Led service quality - practices

Human-Led Services

Qualities
- Reliability
- Responsiveness
- Trustworthiness
- Empathy
- Attractiveness

Valerie Zeithaml and Mary Jo Bitner created the most widely used and accepted model for professional service quality, in 5 dimensions:

Dimension	Definition
1. Reliability	Ability to perform the promised service dependably and accurately
2. Responsiveness	Willingness to help customers and to provide prompt service
3. Trustworthiness	Knowledge and courtesy of employees, and their ability to convey trust and confidence
4. Empathy	Provision of caring, individualized attention to customer
5. Attractiveness	Tangible appearance of physical facilities, equipment, personnel and communication materials

Figure 04-02-03.2 Five dimensions of Human-Led services qualities   Source: Adapted from SERVQUAL
04-02-03. Service Qualities – Human-Led Services - Definition, desired state, best practices

Valerie Zeithaml and Mary Jo Bitner created the most widely used and accepted model for professional service quality, in 5 dimensions shown here. OSM adopts and adapts this model, renaming "Assurance" trustworthiness, and "Tangibles" attractiveness.

As you can see, Human-Led service qualities are in some cases the same (reliability, responsiveness, trustworthiness, and attractiveness) as many of the qualities we seek in an IT-Led services. If for example, you are using an online banking service, you'll want it to be reliable, responsive, trustworthy, and attractive enough. And while an online banking service can't necessarily exhibit empathy per se, the empathy of it's designers for you as the customer or user should shine through in its design.

Human-Led Services

Qualities
- Reliability
- Responsiveness
- Trustworthiness
- Empathy
- Attractiveness

- SERVQUAL Valerie Zeithaml and Mary Jo Bitner
- The Nordstrom Way
- How to Buy/Sell Services HBR
- People (David Maister)
- Facilities Micah Solomon

04-02-03. Service Qualities - Human-Led Services - Definition, desired state, best practices
Module 4: Service concepts, desired states and practices

143

An example: Anything by David Maister is great advice for managing Human-Led services. While most of it is pointed towards consulting firms, the advice is sound and impactful nonetheless. Here are two classics:

- Maister, David H. Managing the Professional Service Firm. (New York: Free Press, 1993.) Maister is a former Harvard Business School professor and is currently a consultant to the world's top professional firms. He is recognized as the foremost expert in professional service firm management and this is the first comprehensive text on the managerial problems of professional firms. Maister explores a wide range of topics including marketing, staffing, service quality, and personal development. The book is full of insightful and practical advice.

- Maister, David H. True Professionalism: The Courage to Care About Your People, Your Clients, and Your Career. (New York: Free Press, 1997.) In this follow-on to Managing the Professional Service Firm, Maister discusses his definition of "true professionalism"--a personal commitment to self-betterment and a dedication to providing excellence and efficiency in client service. He also strongly emphasizes the importance of ethical behavior as the primary means to commercial success. Maister examines these subjects at both the individual level and the firm level and includes excellent recommendations.

Another worth of mention is, Wittreich, Warren J. "How to Buy/Sell Professional Services." (Harvard Business Review, March-April 1966.) The author explores the complexity of buying and selling professional services and provides guidance to both buyers and sellers. Among Wittreich's key ideas are that the selling and rendering of a service can seldom be separated and that the majority of the "sale" actually occurs in delivering on the initial promise made when the engagement was initiated.

Human-Led service quality – Reliability – definition

Human-Led Services

Qualities
- Reliability
- Responsiveness
- Trustworthiness
- Empathy
- Attractiveness

Reliability

The demonstrated capability to perform, and track record of performing, the service dependably and accurately, as promised to, and perceived by, customers and users.

Source: Adapted from SERVQUAL
04-02-04. Human-Led service quality – Reliability – Definition, desired state, best practices
Module 4: Service concepts, desired states and practices

144

Your Human-Led services have the quality of being reliable to the extent that customers and users perceive them as being so; this perception comes from either a demonstrated capacity to do so, or through a track record of doing so, in either case directly or by a trusted reference.

Human-Led Services
Qualities
• Reliability
• Responsiveness
• Trustworthiness
• Empathy
• Attractiveness

Performance is effective when...

- We commit to deliver something by a certain time and follow through on that commitment, both in terms of what is delivered, and when
- We perform services right the first time
- Service providers provide the service at the time they commit to doing so
- Service providers keep error-free records
- The service we provide is consistently as it should be, regardless of variable such as the time of day it is delivered, or by what person or team

Adapted from: https://www.ermt.net/docs/papers/Volume_6/1_January2017/V6N1-132.pdf and www.ec.tuwien.ac.at/~dorn/Courses/SDMC/RATER%20Questionnaire.pdf
04-02-04. Human-Led service quality – Reliability – Definition, desired state, best practices

One good way to think about what your customers and users look for in the qualities of your services is to think through some good and bad service experiences you have had, and to write down what was good and bad about them, then ask yourself, "how would my customers and users say my services compare along these lines?"

Human-Led Services

Qualities
- Reliability
- Responsiveness
- Trustworthiness
- Empathy
- Attractiveness

- ■ <u>SERVQUAL</u> Valerie Zeithaml and Mary Jo Bitner
- ■ <u>How to Buy/Sell Services</u> HBR
- ■ <u>Trusted Advisor (David Maister)</u>
- ■ <u>The 7 Habits of Highly Effective People</u>
- ■ <u>TSIA</u>
- ■ <u>Being a more reliable professional</u>

Being reliable is about being true to your word, confronting mistakes, making sure service is done right the first time, and ultimately, consistently under-promising and over-delivering.

The key ingredient here is a service mindset and an understanding of how to execute on it. Some things that help include:

- Scheduling systems

- Committing to wait time

- Record keeping and review, and

- Performance reviews

David Maister's book, The Trusted Advisor, is an excellent reference, as is Stephen Covey's 7 Habits of Highly Effective People. While the first focuses on consulting firms and the latter on individual behaviors, the advice in both cases applies very much to organizations as providers of services.

Human-Led service quality – Responsiveness – definition

Human-Led Services
Qualities
• Reliability
• Responsiveness
• Trustworthiness
• Empathy
• Attractiveness

Responsiveness

The demonstrated capability and willingness to get back to customers and users quickly when they make a request, and listen, respond and help, and track record of get back to customers and users quickly, and listening, responding and helping them with prompt service, as perceived by customers and users.

Customers and users see our Human-Led services as responsive when we do things like getting back to them quickly when they make a request, and providing prompt service.

Performance is effective when...

Human-Led Services
Qualities
• Reliability
• Responsiveness
• Trustworthiness
• Empathy
• Attractiveness

- We get back to customers and user quickly, acknowledging them when they make a request
- We tell customers and users up front, exactly when services will be executed
- We fulfill services to customers and users stakeholders promptly
- When there is an issue with our service, we handle that promptly, too
- We are always willing to provide prompt service and answer customer and user questions
- We are never too busy to respond to stakeholder requests
- Public situations are treated with care and seriousness
- Customers and users find it easy to talk to a knowledgeable service member when they have a request or issue
- Customers and users find it easy to reach the right service provider in person, by telephone, or email or chat, or by whatever channel we have provided and that they find convenient
- Customers and users say that service access points for our services are conveniently located

Adapted from: https://www.ermt.net/docs/papers/Volume_6/1_January2017/V6N1-132.pdf and www.ec.tuwien.ac.at/~dorn/Courses/SDMC/RATER%20Questionnaire.pdf
04-02-05. Human-Led service quality – Responsiveness – Definition, desired state, best practices
Module 4: Service concepts, desired states and practices

148

When you look at this list, you should see that good performance here needs to be underpinned by things like:

- Good tools, for example, for scheduling

- Good processes and procedures, e.g., for exception handling

- Well-trained service personnel, with the right skills, knowledge and mindset

- The right service culture and rewards

- The right services in the first place, and the right mechanisms for setting expectations and perceptions

Human-Led Services

Qualities
- Reliability
- Responsiveness
- Trustworthiness
- Empathy
- Attractiveness

- <u>Hire the right people in the first place</u>
- <u>Moments of Truth</u>
- <u>SERVQUAL</u> Valerie Zeithaml and Mary Jo Bitner
- <u>The Nordstrom Way</u>
- <u>People (David Maister)</u>

04-02-05. Human-Led service quality – Responsiveness – Definition, desired state, best practices

Module 4: Service concepts, desired states and practices

149

When you look at this list, you should see that good performance here needs to be underpinned by things like:

- The right, well-trained service personnel, with the right skills, knowledge and mindset

- Good tools, for example, for scheduling

- Good processes and procedures, e.g., for exception handling

- The right service culture and rewards

- The right services in the first place, and the right mechanisms for setting expectations and perceptions

The first item on this list is most important. Everything starts and ends with hiring the right people in the first place. In practice, other than by parents, instilling a service mindset in a person who just doesn't have that mindset is an exercise in futility.

Another key element is making sure you are focusing on making "moments of truth" shine—Jan Carlzon's book is the classic on such matters, and is actually a cornerstone book in the birth of service management. While later iterations of traditional ITSM guidance features some over-pivot on process reengineering, it is refreshing to recall the roots of service management, lightweight, agile, and focused on people and individual interactions, and making them better.

Human-Led Services	Trustworthiness
Qualities	The ability to convey and provide assurance of trust and confidence in services.
• Reliability • Responsiveness • Trustworthiness • Empathy • Attractiveness	

- ■ Typically achieved through knowledge and courtesy of employees

Source: Adapted from SERVQUAL
04-02-06. Human-Led service quality – Trustworthiness – Definition, desired state, best practices

Module 4: Service concepts, desired states and practices

150

As a consumer, you use many services. You get haircuts, you use a mobile phone service, and so on. How can you trust that you'll get what you're paying for, that the results will be satisfactory?

If it's the first time you're getting you haircut at a place, you'd look for evidence that it's the place for you, that they'll do a good job. For example, you'll check out how people walking out look—how's their hair? If it's not good, you'll move on and find another place. You might also ask a friend for a referral. So in new-to-you situations, you'll look for evidence or trustworthiness.

If you've been using a service for a while, trust is built through repeated delivery on commitments—in this case, a long track record of good haircuts. Should personnel or other factors change, results might change, and trust might be broken after a bad results.

So, flip this situation around and apply it to how you assure trust in your services.

Performance is effective when...

- Our behavior instills confidence in customers and users, so they feel safe in transactions with us, and with our premises and equipment, and they see us as courteous and competent
- We have, and demonstrate that we have, the skills, knowledge and mindset to do the service well and address customer / user needs
- Materials we provide with the service are appropriate / up-to-date
- Customers and users say we use technology to perform the service quickly and skillfully and appear to know what we are doing and that they are confident that we have and will perform services correctly
- We have a good reputation with customers and users
- We do not pressure customers and users
- We respond accurately and consistently to customer / user inquiries
- We guarantee our services
- We store customer and user documents, records and other data secure from unauthorized use

Human-Led Services

Qualities
- Reliability
- Responsiveness
- Trustworthiness
- Empathy
- Attractiveness

Adapted from: https://www.ermt.net/docs/papers/Volume_6/1_January2017/V6N1-132.pdf andwww.ec.tuwien.ac.at/~dorn/Courses/SDMC/RATER%20Questionnaire.pdf
04-02-06. Human-Led service quality – Trustworthiness – Definition, desired state, best practices

Some things that contribute to these desired states:

- Communications, especially keeping confidentiality

- Soft-skills training (courteousness)

- Knowledge/skills/technical training

- Material review and updated regularly

- Guarantee policies

- Safety ensured

- Security

- Record keeping

Human-Led Services

Qualities
- Reliability
- Responsiveness
- Trustworthiness
- Empathy
- Attractiveness

- ■ <u>SERVQUAL</u> Valerie Zeithaml and Mary Jo Bitner
- ■ <u>People (David Maister)</u>
- ■ <u>Trusted Advisor (David Maister)</u>
- ■ <u>The 7 Habits of Highly Effective People</u>

04-02-06. Human-Led service quality - Trustworthiness - Definition, desired state, best practices

Module 4: Service concepts, desired states and practices

152

David Maister's book, The Trusted Advisor, is an excellent reference, as is Stephen Covey's 7 Habits of Highly Effective People. While the first focuses on consulting firms and the latter on individual behaviors, the advice in both cases applies very much to organizations as providers of services.

Other practices that can contribute to trustworthiness as a service provider include:

- Staff training: trusted advisor skills, service provision skills, knowledge and mindset—consistent across staff

- Safety and security for transactions, data, premises, and equipment

- Mechanisms to ensure service materials are appropriate and current

- Demonstration of competence, e.g., displaying of certifications, endorsements

Human-Led Services
Qualities
• Reliability
• Responsiveness
• Trustworthiness
• Empathy
• Attractiveness

Empathy

Ability to listen and understand the customer and user situation and needs, and to convey caring, individualized attention in the provision of service.

Source: Adapted from SERVQUAL
04-02-07. Human-Led service quality – Empathy – Definition, desired state, best practices
Module 4: Service concepts, desired states and practices

153

Think about service situations you've had recently—maybe it was rescheduling a flight with an airline, perhaps working with a bank on an online banking issues for your checking account. Or maybe you had an appliance break down and you were calling the manufacturer.

Did you feel the person on the other end of the phone or chat was just going through the motions, or were they really trying to listen and understand what your situation was, and what you needed? In other words, was their posture, "I am here to help you with your problem"?

Now flip that around for the services you provide. What do your customers and users think of you? Do you know if they find you empathetic? You may be good to go, or you may have some work to do in this area—the first step is knowing their perceptions and expectations.

Human-Led service quality – Empathy – desired state

Human-Led Services

Qualities
- Reliability
- Responsiveness
- Trustworthiness
- Empathy ◀
- Attractiveness

Performance is effective when...
- We give customers / users individualized attention, convenient hours
- Customers and users say we have their best interest at heart and work to understand their specific needs and objectives
- We recognize each regular customer / user and address them by name
- Our level and cost of service is consistent with what customers and users need and can afford
- We are polite, respectful, considerate and friendly, with a pleasant demeanor, and refrain from acting busy or being rude, and respond politely and with consideration when customers / users ask questions
- We listen to customer and users and show understanding and concern
- We can explain clearly various options available to a particular query
- We avoid using technical jargon when speaking with customers / users
- We contact customers / users if we will miss a scheduled appointment

Adapted from: https://www.ermt.net/docs/papers/Volume_6/1_January2017/V6N1-132.pdf and
www.ec.tuwien.ac.at/~dorn/Courses/SDMC/RATER%20Questionnaire.pdf
04-02-07. Human-Led service quality – Empathy – Definition, desired state, best practices

Empathy can be summed up as the feeling that the service provider has the attitude, "I am here to help you with your request or issue". It's about listening, and social and customer service skills as much as anything else. The opposite of an empathetic person is one who give off the attitude, "I am here to demonstrate my technical skills—bow down, you're not worthy!".

As with other Human-Led service qualities, it's important to understand how the quality of empathy is perceived by your customers and users—it's not what you think that counts.

Human-Led Services
Qualities
• Reliability
• Responsiveness
• Trustworthiness
• Empathy ◀
• Attractiveness

- **SERVQUAL** Valerie Zeithaml and Mary Jo Bitner
- The Nordstrom Way
- Developing Empathy
- The 7 Habits of Highly Effective People

You can avoid a lot of hassles with services staff by hiring people with a service orientation in the first place. Yes, they must have technical skills to succeed, but without an attitude of, "I am here to help you with your request or issue", all will be lost.

Some people contend that you can train people into a service mindset. This may be so, but often it seems that the only ones that can do that is a person's parents. By the time they get to the workplace, it may be an intractable situation for some people, so it's best to hire for a service skill and mindset from the start.

The 7 Habits, especially, "see first to understand, then to be understood", are an excellent place to start in developing empathy.

Some other ways to developing or showing empathy include:

- Training for customer service skills, knowledge and mindset, including other-centered listening and communication skills

- Using customers and users names when dealing with them

- Acting on, or forwarding along for action, customer and user feedback

Human-Led service qualities - Attractiveness – definition

Human-Led Services
Qualities
• Reliability
• Responsiveness
• Trustworthiness
• Empathy
• Attractiveness ◄

Attractiveness

Ability to please customers through the tangible appearance of facilities, equipment, personnel, and communication materials.

Source: Adapted from SERVQUAL
04-02-08. Human-Led service quality – Attractiveness – Definition, desired state, best practices
Module 4: Service concepts, desired states and practices

156

Think about the last time you were put off by some aspect of the appearance of a service provider. Maybe you were in a restaurant and the windows and tabletops and restrooms were dirty looking. Maybe you were at a mobile phone shop and the store looked out of date, shopworn, with some broken shelves, chipped paint here and there, and boxes and merchandise kind of messily laying about. Or maybe it was the person helping you—she had pizza stains on her shirt, or was disheveled in her appearance.

These may seem like little things, but they can add up quickly to a customer or user going "tilt", so it's important to pay attention to them.

So, flip this around and think about this for your services. If you provide deskside support, are your personnel neatly dressed, or are they the crew that can't show up with an unwrinkled shirt, or can't keep their pants tucked in? If you provide, for example, concierge repair and update services for client devices at a physical location, how does it look? You may be good to go, or you may have some work to do. Be honest, and make sure you're ship shape.

Human-Led service quality - Attractiveness - desired state

Human-Led Services
Qualities
• Reliability
• Responsiveness
• Trustworthiness
• Empathy
• Attractiveness

Performance is effective when...

- The tangible aspects of our services—what customers and users can see and touch—are pleasing to customers and users
- Customers and users find our materials associated with the service (pamphlets or statements) visually pleasing and easy to understand
- Customers and user find appearance of service providers pleasing; we dress appropriately and have good hygiene
- Customers and users find our physical facilities / fixtures pleasing
- Customers and users say our technology/equipment looks modern

Adapted from: https://www.ermt.net/docs/papers/Volume_6/1_January2017/V6N1-132.pdf and www.ec.tuwien.ac.at/~dorn/Courses/SDMC/RATER%20Questionnaire.pdf
04-02-08. Human-Led service quality - Attractiveness - Definition, desired state, best practices

So when asked, what do customers and users say about the tangible aspects of our services? Do they see them as appealing? Off-putting? We need to know, or we risk losing customers and users, or at least losing points with customer and user satisfaction.

Human-Led Services
Qualities
• Reliability
• Responsiveness
• Trustworthiness
• Empathy
• Attractiveness ◄

- ■ <u>SERVQUAL</u> Valerie Zeithaml and Mary Jo Bitner
- ■ <u>The Nordstrom Way</u>
- ■ <u>Facilities</u> Micah Solomon
- ■ The Nordstrom Way
- ■ <u>Customer Experience Management (CXM)</u>
- ■ <u>User Experience Management (UXM)</u>

Micah Solomon's writing on retail facilities are a good place to start on understanding the impact of the tangible on customer and user perceptions.

Think about a time where you were super-pleased with a service encounter in a physical space—what made it great?

Now think of another encounter were you were really not happy—what made it lousy for you?

Lastly, think about your services and their tangible aspects. How do they compare against these two scenarios?

You may have this covered, or you may have some work to do.

Attention needs to be paid here to:

- Physical facilities and fixtures
- Product and material appearance, layout and organization
- Staff appearance—dress and hygiene
- Technology appearance

CXM, an emerging field for IT-Led systems, and UXM, are in some ways the parallel best practices for attractiveness in Human-Led systems, and can be food for thought for the physical experience.

IT-Led service quality – definition

IT-Led Services

Services driven by information technology that may be assisted or facilitated by humans, e.g., Office365. Contrast this with Human-Led services, which are driven by humans, and may be assisted or facilitated by IT, for example, PC repair, Moves/Adds/Changes.

IT-Led Service Qualities

How an IT-Led service behaves, it's non-functional characteristics, including its availability, trustworthiness, usability and adaptability, and their associated characteristics.

IT-Led Services	
Qualities	
Availability	Usability
• Manageability	• Internationalization
• Serviceability	• Accessibility
• Performance	Adaptability
• Reliability	• Interoperability
• Recoverability	• Scalability
• Discoverability	• Elasticity
Trustworthiness	• Portability
• Security	• Extensibility
• Integrity	
• Credibility	
• Compliance	

Figure 04-02-09.1 IT-Led services qualities
04-02-09. Service Qualities – IT-Led Services – Definition, desired state, best practices

Module 4: Service concepts, desired states and practices

IT-Led service qualities include:
- Availability
- Manageability
- Serviceability
- Performance
- Reliability
- Recoverability
- Discoverability
- Trustworthiness
- Security
- Integrity
- Credibility
- Compliance
- Usability
- Internationalization
- Accessibility
- Adaptability
- Interoperability
- Scalability
- Elasticity
- Portability
- Extensibility

While this is a long list, it is by no means comprehensive or final. IT-Led service qualities are basically any non-functional characteristic of a service.

IT-Led service quality – desired state

Performance is effective when...

- Customers and users and other stakeholder (e.g., those concerned with governance / regulatory compliance, or accessibility, or security, or internationalization, disaster recovery, and so on), say our services have the right set of qualities, including availability, manageability, serviceability, performance, reliability, recoverability, discoverability, trustworthiness, security, integrity, credibility, compliance, usability, internationalization, accessibility, adaptability, interoperability, scalability, elasticity, portability, and extensibility)

IT-Led Services	
Qualities	
Availability	Usability
• Manageability	• Internationalization
• Serviceability	• Accessibility
• Performance	Adaptability
• Reliability	• Interoperability
• Recoverability	• Scalability
• Discoverability	• Elasticity
Trustworthiness	• Portability
• Security	• Extensibility
• Integrity	
• Credibility	
• Compliance	

04-02-09. Service Qualities – IT-Led Services – Definition, desired state, best practices

Module 4: Service concepts, desired states and practices

160

When you look at this list of qualities, you should see a number that pop out as important to customers and users.

The same goes for the provider.

Suppliers, also, are stakeholders in this, for the IT-Led services they provide and contribute to.

And besides the four primary stakeholders of service management—customers, users, the provider, and supplier—there are others that have a stake in these qualities. Those that own regulatory compliance will be concerned about ensuring compliance with, e.g., PCI, Sox, Basel 2, and SAS 70, or whatever regulations are pertinent. Those that own security will want to make sure the services are secure. And those concerned with uptake of the service will want to make sure perhaps that it is 401 compliant for accessibility, or supports French, Dutch, Spanish, etc. where those languages are needed.

IT-Led service quality – practices

- Service Leve Management
- Service Level / Operational Level Agreements
- Underpinning Contracts
- Non-functional requirements

IT-Led Services	
Qualities	
Availability	Usability
• Manageability	• Internationalization
• Serviceability	• Accessibility
• Performance	Adaptability
• Reliability	• Interoperability
• Recoverability	• Scalability
• Discoverability	• Elasticity
Trustworthiness	• Portability
• Security	• Extensibility
• Integrity	
• Credibility	
• Compliance	

Two additional sources of best practice are:

https://dalbanger.wordpress.com/2014/01/08/a-basic-non-functional-requirements-checklist/, and

carrotschool.com/blog/the-definitive-non-functional-requirements-checklist

IT-Led service quality - Availability - definition

Availability

Ability of a service or service component to perform its required function at an agreed instant or over an agreed period. NOTE Availability is normally expressed as a ratio or percentage of the time that the service or service component is actually available for use by the customer to the agreed time that the service should be available.

- Reliability & Maintainability - see definitions here https://www.iso.org/obp/ui/#iso:std:iso-iec:2382:-14:ed-2:v1:en
- Reliability measured by MTBF; maintainability by MTTR

IT-Led Services	
Qualities	
Availability	Usability
• Manageability	• Internationalization
• Serviceability	• Accessibility
• Performance	Adaptability
• Reliability	• Interoperability
• Recoverability	• Scalability
• Discoverability	• Elasticity
Trustworthiness	• Portability
• Security	• Extensibility
• Integrity	
• Credibility	
• Compliance	

Source: ISO/IEC 20000-1:2011(E) 3 Terms and definitions
04-02-10. IT-Led service quality - Availability - Definition, desired state, best practices

Module 4: Service concepts, desired states and practices

162

Availability is percentage of time within the agreed or committed window a service is up. It's made up of reliability, which is the "keep it up and running part", as measured by meant time between failures or MTBF, and maintainability, an engineering term that means the ease with which a service which has failed can be returned to working order, which is measured by mean time to repair, or MTTR.

It's important to note that availability is the quality related to NORMAL operating conditions, e.g., it's a regular Tuesday here at XYZ corporation, whereas Recoverability is the quality related to the ABNORMAL situation of being out of business—and regaining availability of services in priority order to get back into business. While measures taken for the one can benefit the other, and vice versa, they are two distinct domains and qualities of services.

IT-Led service quality – Availability – desired state

Performance is effective when...

- We ensure availability of services in normal business operations by designing and managing services to keep them up and running. Should they go down, get them up quickly within specified service levels and costs
- We develop services for the required levels of availability in the first place, with some "wiggle room" so that we can consistently hit targets; this includes design for reliability (keep thing up and running, measured by MTBF), maintainability (if they go down, get them up quick, measured by MTTR)
- Services provide us with easily accessible, accurate data on their availability
- We work to take measures and choose platforms that provide the levels of automated failover and recovery our stakeholders need; to the extent possible on the platforms we are on, we automate failover and recovery

IT-Led Services	
Qualities	
Availability	Usability
• Manageability	• Internationalization
• Serviceability	• Accessibility
• Performance	Adaptability
• Reliability	• Interoperability
• Recoverability	• Scalability
• Discoverability	• Elasticity
Trustworthiness	• Portability
• Security	• Extensibility
• Integrity	
• Credibility	
• Compliance	

04-02-10. IT-Led service quality – Availability – Definition, desired state, best practices

Design for both reliability and maintainability are required to promote the right levels of availability; one or the other does not cut it. Some IT service provider shops are very engineering oriented towards reliability, and stumble when things go down because they don't usually go down. Others are really good at getting things back up and running, because they have a lot of practice, because the engineering for reliability just isn't there. Balance is needed here.

- Design for reliability – Building Dependable Systems
- Design for maintainability Michael Tortorella
- Cloud reliability – Simian Army / Antifragility
- Cloud maintainability
- Availability management
- Availability levels
- Public status pages / transparent uptime
- Incident Command System (ICS)
- Kepner Tregoe problem analysis

IT-Led Services	
Qualities	
Availability	Usability
• Manageability	• Internationalization
• Serviceability	• Accessibility
• Performance	Adaptability
• Reliability	• Interoperability
• Recoverability	• Scalability
• Discoverability	• Elasticity
Trustworthiness	• Portability
• Security	• Extensibility
• Integrity	
• Credibility	
• Compliance	

04-02-10. IT-Led service quality - Availability - Definition, desired state, best practices
Module 4: Service concepts, desired states and practices

164

For reliability, even though it's ancient (1994), the Digital Equipment Corporation publication, "Building Dependable Systems" remains a very clear and well written guide that can be applied to any service.

A key part of availability is getting availability levels right, because each involve a price / performance tradeoff. These levels include:

- High availability. The service is available during specified operating hours with no unplanned outages.

- Continuous operations. The service is available 24 hours a day, 7 days a week, with no scheduled outages.

- Continuous availability. The service is available 24 hours a day, 7 days a week, with no planned or unplanned outages, for example, like amazon.com.

Public status pages are a great way to build confidence in you "skin in the game" on availability.

And while it's really a way of handling major incidents, ICS deserves mention here because of it's potential for reducing MTTR in reactive situations. The same goes for KT analysis.

IT-Led service quality – Manageability – definition

Manageability

The ability to gather information about the state of something and to control it.

- Manageability is largely a function of the quality and extent of service instrumentation, telemetry and logging, and automation for remedial action.

IT-Led Services	
Qualities	
Availability	Usability
▷ Manageability	• Internationalization
• Serviceability	• Accessibility
• Performance	Adaptability
• Reliability	• Interoperability
• Recoverability	• Scalability
• Discoverability	• Elasticity
Trustworthiness	• Portability
• Security	• Extensibility
• Integrity	
• Credibility	
• Compliance	

Source: www.opengroup.org/public/arch/p3/trm/tx/tx_quals.htm
04-02-11. IT-Led service quality – Manageability – Definition, desired state, best practices

Module 4: Service concepts, desired states and practices

Instrumentation refers to an ability to monitor or measure the level of a service's performance, to diagnose errors and to write trace information. Developers implement instrumentation in the form of code instructions that monitor specific components in a service (for example, instructions may output logging information to appear on screen). When a service contains instrumentation code, it can be managed using a management tool. Instrumentation is necessary to review the performance of the service. Source: adapted from https://en.wikipedia.org/wiki/Instrumentation_(computer_programming)

Telemetry is a term for technologies that accommodate collecting information in the form of measurements or statistical data, and forward it to IT systems in a remote location. This term can be used in reference to many different types of systems, such as wireless systems using radio, ultrasonic or infrared technologies, or some types of systems operating over telephone or computer networks. Others may use different strategies like SMS messaging. Source: https://www.techopedia.com/definition/14853/telemetry

IT-Led service quality – Manageability – desired state

Performance is effective when...

- We can easily and automatically gather information on the state of all things worth managing—stakeholders, services, and the SMS—compare it to desired states, and take action where there is a gap
- We generally don't have gaps in manageability, but when we do, we close them quickly and permanently

IT-Led Services	
Qualities	
Availability	Usability
Manageability	• Internationalization
• Serviceability	• Accessibility
• Performance	Adaptability
• Reliability	• Interoperability
• Recoverability	• Scalability
• Discoverability	• Elasticity
Trustworthiness	• Portability
• Security	• Extensibility
• Integrity	
• Credibility	
• Compliance	

Manageability must be designed for and continuously improved for it to be effective for services. Instrumentation, telemetry, and logging and monitoring functionality don't just appear out of nowhere—a proper health model should ship with a service, as part of the design.

IT-Led service quality – Manageability – practices

- <u>Making it manageable</u> fuel Palo Alto
- <u>Instrumentation / Telemetry</u>
- <u>Monitoring</u>
- <u>Cloud Monitoring / Telemetry</u>
- <u>Logging / Instrumentation</u>
- <u>Logging</u> cheat sheet
- <u>Event management</u>

IT-Led Services	
Qualities	
Availability	Usability
→ Manageability	• Internationalization
• Serviceability	• Accessibility
• Performance	Adaptability
• Reliability	• Interoperability
• Recoverability	• Scalability
• Discoverability	• Elasticity
Trustworthiness	• Portability
• Security	• Extensibility
• Integrity	
• Credibility	
• Compliance	

While event management / service monitoring and control are part of the service management system, they are joined at the hip with the quality of manageability, so they are mentioned here. In the end, manageability is about putting hooks into services so they can be monitored and controlled.

IT-Led service quality - Serviceability - definition

Serviceability

The ability to identify problems and take corrective action such as to repair or upgrade a component in a running system

- Services are serviceable to the extent that, in a repair, maintenance, or restocking situations, what to do, where, and how and so on is readily apprehended and easily reached

IT-Led Services	
Qualities	
Availability	Usability
• Manageability	• Internationalization
Serviceability	• Accessibility
• Performance	Adaptability
• Reliability	• Interoperability
• Recoverability	• Scalability
• Discoverability	• Elasticity
Trustworthiness	• Portability
• Security	• Extensibility
• Integrity	
• Credibility	
• Compliance	

Source: www.opengroup.org/public/arch/p3/trm/tx/tx_quals.htm
https://en.wikipedia.org/wiki/Serviceability_(computer)
04-02-12. IT-Led service quality - Serviceability - Definition, desired state, best practices

Module 4: Service concepts, desired states and practices

One difference between OSM and traditional ITSM guidance should be clear here. Traditional ITSM guidance uses the term, "serviceability" in a way that is out of step with the industry, attaching it to a supplier's contribution to the uptime or downtime of a service or one of it's components. The typical definition, which is taken up here by the Open Group, is basically—how easy is it to service?

IT-Led service quality – Serviceability – desired state

Performance is effective when:

- Our services are designed to help identify problems and take corrective action, such as to repair or upgrade to a service or one of its components
- In repair, maintenance, and restocking situations, what to do, where, and how and so on is readily apprehended and easily reached

IT-Led Services	
Qualities	
Availability	Usability
• Manageability	• Internationalization
Serviceability	• Accessibility
• Performance	Adaptability
• Reliability	• Interoperability
• Recoverability	• Scalability
• Discoverability	• Elasticity
Trustworthiness	• Portability
• Security	• Extensibility
• Integrity	
• Credibility	
• Compliance	

www.opengroup.org/public/arch/p3/trm/tx/tx_quals.htm
04-02-12. IT-Led service quality – Serviceability – Definition, desired state, best practices

For example, if I am to work on an office printer / copier / scanner with a tray feeding system, how hard or easy is it to clear jams, to figure out which tray is what, and so on, while trying to service the unit? How hard or easy is maintenance, or to replenish the paper or toner? Is it ready to hand? That is the quality of being serviceable. It requires design for serviceability, as it's much harder and costlier to try to "bolt it on" later.

IT-Led service quality - Serviceability - practices

- **Design for serviceability** NDP
- **Design for serviceability** Guy Carafone
- **Design for supportability**

IT-Led Services	
Qualities	
Availability	Usability
• Manageability	• Internationalization
Serviceability	• Accessibility
• Performance	Adaptability
• Reliability	• Interoperability
• Recoverability	• Scalability
• Discoverability	• Elasticity
Trustworthiness	• Portability
• Security	• Extensibility
• Integrity	
• Credibility	
• Compliance	

NDP lists he key design for serviceability guidelines as:

- Simplification

- Standardization

- Access

- Ergonomics

- Safety

- Disconnecting/Reconnecting

- Unfastening/Refastening

- Part Handling

- Location and Insertion, and

- Mistake-Proofing

Performance

The ability of a component to perform its tasks in an appropriate time. Source: www.opengroup.org/public/arch/p3/trm/tx/tx_quals.htm

Throughput

A measure of how many units of information a service can process in a given amount of time. It is applied broadly to services ranging from various aspects of computer and network systems to organizations. Related measures of service productivity include the speed with which some specific workload can be completed, and response time, the amount of time between a single interactive user request and receipt of the response.

IT-Led Services	
Qualities	
Availability	Usability
• Manageability	• Internationalization
• Serviceability	• Accessibility
Performance	Adaptability
• Reliability	• Interoperability
• Recoverability	• Scalability
• Discoverability	• Elasticity
Trustworthiness	• Portability
• Security	• Extensibility
• Integrity	
• Credibility	
• Compliance	

Performance, or throughput, is a function of capacity, of, e.g., storage, compute, and network.

Traditional ITSM guidance focuses on capacity, which is a means to the end; the end is performance. OSM focuses on performance.

IT-Led service quality – Performance – desired state

Performance is effective when...
- Performance meets committed rates for our services
- We ensure capacity matches demand at the right time and at the right cost by seeking to understand current and future demand and capacity, and delivering the right resources, at the right time and at the right cost. We understand and influence demand, and avoid excess and insufficient capacity and associated costs and service impact.
- We design services for required performance levels up front, with "wiggle room" so we consistently hit targets; this includes design and automation for elasticity—scaling to meet demand, and with easily accessible, accurate performance data
- When we size services, we factor in consumers' patterns of business activity where possible

IT-Led Services	
Qualities	
Availability	Usability
• Manageability	• Internationalization
• Serviceability	• Accessibility
Performance	Adaptability
• Reliability	• Interoperability
• Recoverability	• Scalability
• Discoverability	• Elasticity
Trustworthiness	• Portability
• Security	• Extensibility
• Integrity	
• Credibility	
• Compliance	

04-02-13. IT-Led service quality - Performance - Definition, desired state, best practices

This article https://en.wikipedia.org/wiki/Computer_performance does a good job of laying out aspects of performance for computers that can apply also to networks, virtual machines, etc.

- **Capacity management**
- **Demand management**
- **Load balancing**
- **Network performance management**
- **Database performance management** SQL
- **Autoscale** scale up/down in/out
- **Elasticity**

IT-Led Services	
Qualities	
Availability	Usability
• Manageability	• Internationalization
• Serviceability	• Accessibility
Performance	Adaptability
• Reliability	• Interoperability
• Recoverability	• Scalability
• Discoverability	• Elasticity
Trustworthiness	• Portability
• Security	• Extensibility
• Integrity	
• Credibility	
• Compliance	

04-02-13. IT-Led service quality - Performance - Definition, desired state, best practices
Module 4: Service concepts, desired states and practices

173

There are many more areas of performance management, this is just a sampling. Some other practices to enhance performance include:

- Automation
- Trend analysis
- Tuning
- Influencing demand (e.g., by differential charging)
- Workload analysis
- Performance analysis / benchmarking

Reliability

The ability to resist failure

Source: Adapted from
www.opengroup.org/public/arch/p3/trm/tx/tx_quals.htm

IT-Led Services	
Qualities	
Availability	Usability
• Manageability	• Internationalization
• Serviceability	• Accessibility
• Performance	Adaptability
→ Reliability	• Interoperability
• Recoverability	• Scalability
• Discoverability	• Elasticity
Trustworthiness	• Portability
• Security	• Extensibility
• Integrity	
• Credibility	
• Compliance	

Source: Adapted from www.opengroup.org/public/arch/p3/trm/tx/tx_quals.htm
04-02-14. IT-Led service quality - Reliability - Definition, desired state, best practices
Module 4: Service concepts, desired states and practices

174

Reliability is the extent to which a service stays up and running, and is measure by mean time between failures (MTBF).

IT-Led service quality - Reliability - desired state

Performance is effective when...

- Generally, services don't fail, because of how we plan, build, deploy, test and release them
- Should services fail, we find out why and harden the service(s) so failure does not re-occur

IT-Led Services	
Qualities	
Availability	Usability
• Manageability	• Internationalization
• Serviceability	• Accessibility
• Performance	Adaptability
Reliability	• Interoperability
• Recoverability	• Scalability
• Discoverability	• Elasticity
Trustworthiness	• Portability
• Security	• Extensibility
• Integrity	
• Credibility	
• Compliance	

As with many similar qualities, reliability has to be designed into a service up front, and continually improved, in this case to support service objectives or commitments for uptime.

IT-Led service quality - Reliability - practices

- Design for reliability – Building Dependable Systems
- Design for reliability – DFR including FMEA
- Fault tree analysis
- Cloud reliability – Simian Army / Antifragility
- Availability management
- Availability levels'
- Server hardening

IT-Led Services	
Qualities	
Availability	Usability
• Manageability	• Internationalization
• Serviceability	• Accessibility
• Performance	Adaptability
Reliability	• Interoperability
• Recoverability	• Scalability
• Discoverability	• Elasticity
Trustworthiness	• Portability
• Security	• Extensibility
• Integrity	
• Credibility	
• Compliance	

04-02-14. IT-Led service quality - Reliability - Definition, desired state, best practices
Module 4: Service concepts, desired states and practices
176

A number of techniques are available for designing more reliable systems, including Failure Modes and Effect Analysis (FMEA) is a step-by-step approach for identifying all possible ways, or modes, in which a service might fail, and the effects of each failure, as a basis for prioritizing hardening activities.

IT-Led service quality – Recoverability – definition

Recoverability

Ability to restore a service to a known good working state after a failure has occurred, through techniques like backups, baseline images and snapshots, hardware and software records, redundant failover equipment, and reference records and code through which a known good state can be restored.

Source: adapted from
opengroup.org/public/arch/p3/trm/tx/tx_quals.htm

IT-Led Services	
Qualities	
Availability	Usability
• Manageability	• Internationalization
• Serviceability	• Accessibility
• Performance	Adaptability
• Reliability	• Interoperability
Recoverability	• Scalability
• Discoverability	• Elasticity
Trustworthiness	• Portability
• Security	• Extensibility
• Integrity	
• Credibility	
• Compliance	

Recoverability requires a restore point—something to rely on to go back to a known good state. It could be, for example, that you're using immutable deployment, where all components are "vanilla", and you simply replace rather than repair or change to restore a known good state. Or, for example, you could use a VM baseline image and apply a snapshot to restore it to what it looked like when the snapshot was taken.

Ensuring you have a good restore point for key components of services should start in design. At what interval will you snapshot? Will you simply record basic information with which you can recreate a component, or a full image? These decisions belong in design. Without them, you'll be left with a need to restore and no way to get back to that known, good, state.

IT-Led service quality - Recoverability - desired state

Performance is effective when...

- We design for recoverability
- We capture known good recovery points for our services and their components, at an interval, and with a technique suitable for restoring services and their component to a known good state in a timely fashion, e.g., through backups, baselines, snapshots, records and code through which known good states can be restored, and redundant failover equipment, etc.
- We test our recovery capability for services thoroughly enough and at a frequent enough intervals, and make improvements based on those tests, such that we are confident in the ability to recover quickly and completely should failures occur

IT-Led Services	
Qualities	
Availability	Usability
• Manageability	• Internationalization
• Serviceability	• Accessibility
• Performance	Adaptability
• Reliability	• Interoperability
Recoverability	• Scalability
• Discoverability	• Elasticity
Trustworthiness	• Portability
• Security	• Extensibility
• Integrity	
• Credibility	
• Compliance	

Good recovery capability requires design up front, ongoing execution of restore point captures, including deltas, and frequent and thorough enough testing and improvement to ensure recovery capability is sufficient.

IT-Led service quality – Recoverability – practices

- <u>Availability Management</u>
- <u>Backup and recovery</u>
- <u>Backups</u> (website)
- <u>Backups</u> (VMs, Hyper-V)

IT-Led Services	
Qualities	
Availability	Usability
• Manageability	• Internationalization
• Serviceability	• Accessibility
• Performance	Adaptability
• Reliability	• Interoperability
▸ Recoverability	• Scalability
• Discoverability	• Elasticity
Trustworthiness	• Portability
• Security	• Extensibility
• Integrity	
• Credibility	
• Compliance	

The general idea here is to capture points you can restore to, in a format that restores the service or component in a timely enough fashion for the situation at hand.

IT-Led service quality - Discoverability - definition

Discoverability

The ability of a system to be located when needed, for example, as in SOA, or in a microservices architecture, or a REST API, or a server that's failed and has been replaced by another that provides the same server, but has, e.g., a different IP address, GUID, etc.

Source: adapted from
www.opengroup.org/public/arch/p3/trm/tx/tx_quals.htm

IT-Led Services	
Qualities	
Availability	Usability
• Manageability	• Internationalization
• Serviceability	• Accessibility
• Performance	Adaptability
• Reliability	• Interoperability
• Recoverability	• Scalability
Discoverability	• Elasticity
Trustworthiness	• Portability
• Security	• Extensibility
• Integrity	
• Credibility	
• Compliance	

04-02-16. IT-Led service quality - Discoverability - Definition, desired state, best practices
Module 4: Service concepts, desired states and practices

180

Services need to be locatable when needed. This is key in service-oriented architecture, where a particular instance goes away, and is replaced by another in a recovery or scaling situation.

Performance is effective when...

■ Services can be easily located when needed

IT-Led Services	
Qualities	
Availability	Usability
• Manageability	• Internationalization
• Serviceability	• Accessibility
• Performance	Adaptability
• Reliability	• Interoperability
• Recoverability	• Scalability
Discoverability	• Elasticity
Trustworthiness	• Portability
• Security	• Extensibility
• Integrity	
• Credibility	
• Compliance	

In a relatively static system, discovery can almost be an afterthought. But in a dynamic environment, as in today's cloud platforms, it is an essential part of locating changing components that provide the same functionality, as they are needed.

IT-Led service quality - Discoverability - practices

- <u>Service Discovery</u> (microservices architecture)
- <u>Service Discovery</u> (and dependency injection)
- <u>Service Discovery</u> (REST API)
- <u>Service Discovery</u> (service oriented architecture)

IT-Led Services	
Qualities	
Availability	Usability
• Manageability	• Internationalization
• Serviceability	• Accessibility
• Performance	Adaptability
• Reliability	• Interoperability
• Recoverability	• Scalability
Discoverability	• Elasticity
Trustworthiness	• Portability
• Security	• Extensibility
• Integrity	
• Credibility	
• Compliance	

Service discovery allow us to not have to know up front, in a static way, where the services that do functions x, y, and z that they depend on are located up front; such a static mapping would create an unnecessary dependency, and quickly breaks down in a dynamic environment.

Trustworthiness

The ability to test and prove / provide evidence and assurance that the service has the security, integrity and credibility required, where security is protection of information from unauthorized access, integrity is the assurance that data has not been corrupted, and credibility is the level of trust in the integrity of the system and its data.
Source: Adapted from www.opengroup.org/public/arch/p3/trm/tx/tx_quals.htm

IT-Led Services
Qualities

Availability
- Manageability
- Serviceability
- Performance
- Reliability
- Recoverability
- Discoverability

Trustworthiness
- Security
- Integrity
- Credibility
- Compliance

Usability
- Internationalization
- Accessibility

Adaptability
- Interoperability
- Scalability
- Elasticity
- Portability
- Extensibility

04-02-17. Service qualities - Trustworthiness - Definition, desired state, best practices
Module 4: Service concepts, desired states and practices

183

Another aspect of trustworthiness is availability, which is one of the three primary tenets of information security, and in this context means that information assets are available to those that are authorized to use them, that is, there is no denial of service to authorized users.

IT-Led service quality – Trustworthiness – desired state

Performance is effective when...

- Services and their data can be and have been tested continuously to demonstrate that they have required levels of:
    - Security: information assets are protected from unauthorized access
    - Integrity: assurance that information assets have not been corrupted
    - Credibility: the level of trust in the integrity of the information assets is high
    - Compliance: it is compliant with all relevant regulatory requirements

IT-Led Services	
Qualities	
Availability	Usability
• Manageability	• Internationalization
• Serviceability	• Accessibility
• Performance	Adaptability
• Reliability	• Interoperability
• Recoverability	• Scalability
• Discoverability	• Elasticity
Trustworthiness	• Portability
• Security	• Extensibility
• Integrity	
• Credibility	
• Compliance	

04-02-17. Service qualities – Trustworthiness – Definition, desired state, best practices
Module 4: Service concepts, desired states and practices

184

See whatis.techtarget.com/definition/Confidentiality-integrity-and-availability-CIA for a good discussion of the CIA triad, or the three primary tenets of information security.

- <u>Trustworthy computing</u>
- <u>Trustworthy cloud computing</u>

IT-Led Services	
Qualities	
Availability	Usability
• Manageability	• Internationalization
• Serviceability	• Accessibility
• Performance	Adaptability
• Reliability	• Interoperability
• Recoverability	• Scalability
• Discoverability	• Elasticity
Trustworthiness	• Portability
• Security	• Extensibility
• Integrity	
• Credibility	
• Compliance	

04-02-17. Service qualities – Trustworthiness – Definition, desired state, best practices

Module 4: Service concepts, desired states and practices

Services and their data can be and have been tested continuously to demonstrate that they have required levels of:

- Security: information assets are protected from unauthorized access

- Integrity: assurance that information assets have not been corrupted

- Credibility: the level of trust in the integrity of the information assets is high

- Compliance: it is compliant with all relevant regulatory requirements

IT-Led service quality – Security – definition

Security

Ability to protect information from unauthorized access
Source:
www.opengroup.org/public/arch/p3/trm/tx/tx_quals.htm

Information Security

Preservation of confidentiality, integrity and accessibility of information NOTE 1 In addition, other properties such as authenticity, accountability, non-repudiation and reliability can also be involved. NOTE 2 The term "availability" has not been used in this definition because it is a defined term in this part of ISO/IEC 20000 which would not be appropriate for this definition. NOTE 3 Adapted from ISO/IEC 27000:2009. Source: ISO/IEC 20000-1:2011(E) 3 Terms and definitions

IT-Led Services	
Qualities	
Availability	Usability
• Manageability	• Internationalization
• Serviceability	• Accessibility
• Performance	Adaptability
• Reliability	• Interoperability
• Recoverability	• Scalability
• Discoverability	• Elasticity
Trustworthiness	• Portability
Security	• Extensibility
• Integrity	
• Credibility	
• Compliance	

Information Security Incident

Single or a series of unwanted or unexpected information security events that have a significant probability of compromising business operations and threatening information security [ISO/IEC 27000:2009]. Source: ISO/IEC 20000-1:2011(E) 3 Terms and definitions

04-02-18. IT-Led service quality – Security – Definition, desired state, best practices

Module 4: Service concepts, desired states and practices

186

IT-Led service quality – Security – desired state

Performance is effective when...

- We align IT security with business security and ensure security is effectively managed in all service and Service activities and provide a focus for all IT security issues and activities.
- We design services that ensure the confidentiality, integrity, and availability of assets, information, data, and IT services
- Access to services and the service management system are provided only to authorized users
- We monitory security and automate security contingencies where possible and useful for faster response to security situations

IT-Led Services	
Qualities	
Availability	Usability
• Manageability	• Internationalization
• Serviceability	• Accessibility
• Performance	Adaptability
• Reliability	• Interoperability
• Recoverability	• Scalability
• Discoverability	• Elasticity
Trustworthiness	• Portability
Security	• Extensibility
• Integrity	
• Credibility	
• Compliance	

Security has to be agile in order to keep up with the number and variety of threats and vulnerabilities in IT today.

- <u>Information Security Management</u>
- <u>Three primary tenets of information security—confidentiality, integrity, and availability (CIA)</u>
- <u>Identify management</u>
- <u>Security policies</u>
- <u>Security controls</u>
- <u>Security and penetration testing</u>
- <u>Security, governance, validation</u>
- <u>Security automation</u>

IT-Led Services	
Qualities	
Availability	Usability
• Manageability	• Internationalization
• Serviceability	• Accessibility
• Performance	Adaptability
• Reliability	• Interoperability
• Recoverability	• Scalability
• Discoverability	• Elasticity
Trustworthiness	• Portability
Security	• Extensibility
• Integrity	
• Credibility	
• Compliance	

The three primary tenets of information security are confidentiality, integrity, and availability (CIA):

1. Confidentiality – information is available to only those with a right to it, and kept private from those who do not have a right to it

2. Integrity – information is complete, accurate and free from unauthorized modification

3. Availability – information can be accessed by those with a right to it when required (there is no denial of service)

Integrity

The ability to assure services and their data have not been corrupted.

Source: www.opengroup.org/public/arch/p3/trm/tx/tx_quals.htm

IT-Led Services	
Qualities	
Availability	Usability
• Manageability	• Internationalization
• Serviceability	• Accessibility
• Performance	Adaptability
• Reliability	• Interoperability
• Recoverability	• Scalability
• Discoverability	• Elasticity
Trustworthiness	• Portability
• Security	• Extensibility
Integrity	
• Credibility	
• Compliance	

Integrity is the quality of information assets, in this case, the data associated with services, where it is assured to be complete, accurate and free from unauthorized modification.

Performance is effective when...

- We are able to continuously demonstrate that services and their data have not been corrupted
- Data is typically not corrupted; in the rare event that it becomes corrupted, we are okay, because we have taken precautions and can restore to a known good state; we also learn how it became corrupted and take steps to learn and improve so that the corruption does not reoccur

IT-Led Services	
Qualities	
Availability	Usability
• Manageability	• Internationalization
• Serviceability	• Accessibility
• Performance	Adaptability
• Reliability	• Interoperability
• Recoverability	• Scalability
• Discoverability	• Elasticity
Trustworthiness	• Portability
• Security	• Extensibility
Integrity	
• Credibility	
• Compliance	

04-02-19. IT-Led service quality – Integrity – Definition, desired state, best practices

Module 4: Service concepts, desired states and practices

190

Services and their data can become corrupted through:

- Power-related issues

- Improper shutdowns / dismounts

- Hardware issues

- Software issues

- Transmission issues

- Faulty updates, and

- Hacking, including data breaches caused by malware, viruses or malicious internal or external attacks.

IT-Led service quality – Integrity – practices

- **Admin, physical, and technical security**
- **Data integrity**

IT-Led Services	
Qualities	
Availability	Usability
• Manageability	• Internationalization
• Serviceability	• Accessibility
• Performance	Adaptability
• Reliability	• Interoperability
• Recoverability	• Scalability
• Discoverability	• Elasticity
Trustworthiness	• Portability
• Security	• Extensibility
▸ Integrity	
• Credibility	
• Compliance	

Services and their data can be protected from corruption through:

- Power conditioning and backup

- Proper shutdowns / dismounts

- Hardware, software and transmission assurance

- Integrity checks after updates, and

- Mitigations and contingencies against malware, viruses or malicious internal or external attacks.

IT-Led service quality - Credibility - definition

Credibility

The ability to ensure the level of trust in the integrity of the service and its data

Source:
www.opengroup.org/public/arch/p3/trm/tx/tx_quals.htm

IT-Led Services	
Qualities	
Availability	Usability
• Manageability	• Internationalization
• Serviceability	• Accessibility
• Performance	Adaptability
• Reliability	• Interoperability
• Recoverability	• Scalability
• Discoverability	• Elasticity
Trustworthiness	• Portability
• Security	• Extensibility
• Integrity	
Credibility	
• Compliance	

IT-Led service quality – Credibility – desired state

Performance is effective when...

- Stakeholders have a well-founded trust in the level of integrity of services and their data, because we take a proactive approach of demonstrating trustworthy data, and provide evidence of our activities and results of doing so

IT-Led Services	
Qualities	
Availability	Usability
• Manageability	• Internationalization
• Serviceability	• Accessibility
• Performance	Adaptability
• Reliability	• Interoperability
• Recoverability	• Scalability
• Discoverability	• Elasticity
Trustworthiness	• Portability
• Security	• Extensibility
• Integrity	
⟹ Credibility	
• Compliance	

Credibility is the ability to ensure the level of trust in the integrity of the service and its data.

IT-Led service quality – Credibility – practices

■ <u>Data integrity</u>

IT-Led Services	
Qualities	
Availability	Usability
• Manageability	• Internationalization
• Serviceability	• Accessibility
• Performance	Adaptability
• Reliability	• Interoperability
• Recoverability	• Scalability
• Discoverability	• Elasticity
Trustworthiness	• Portability
• Security	• Extensibility
• Integrity	
Credibility	
• Compliance	

See
https://www.fda.gov/downloads/drugs/guidancecomplianceregulatoryinformation/guidances/ucm 495891.pdf for an example of data integrity guidance.

IT-Led service quality – Compliance – definition

Compliance

The ability to demonstrate we have the proper governance in place for compliance with all applicable legal, regulatory and other compliance requirements

Common regulations for compliance, for example in data centers include:

- HIPPA
- PCI DSS
- SAS 70
- SSAE
- SOC 1/2/3
- EU-U.S. Privacy Shield

IT-Led Services	
Qualities	
Availability	Usability
• Manageability	• Internationalization
• Serviceability	• Accessibility
• Performance	Adaptability
• Reliability	• Interoperability
• Recoverability	• Scalability
• Discoverability	• Elasticity
Trustworthiness	• Portability
• Security	• Extensibility
• Integrity	
• Credibility	
Compliance	

Governance, Risk and Compliance, or GRC for short, refers to a company's coordinated strategy for managing the broad issues of corporate governance, enterprise risk management (ERM) and corporate compliance with regard to regulatory requirements. Source: https://www.webopedia.com/TERM/G/grc-governance-risk-compliance.html

IT-Led service quality - Compliance - desired state

Performance is effective when...

- Services conform to all applicable legislative and regulatory requirements
- We are good at determining the cost of compliance for stakeholders, services, and the SMS, including money and resources, and advocating for those costs, and building them in to our services and SMS
- We work proactively to have efficient and effective compliance, to reduce the cost and effort to maintain compliance
- Stakeholders, services, and the SMS are all in compliance with policies and laws, e.g.., software licenses

IT-Led Services	
Qualities	
Availability	Usability
• Manageability	• Internationalization
• Serviceability	• Accessibility
• Performance	Adaptability
• Reliability	• Interoperability
• Recoverability	• Scalability
• Discoverability	• Elasticity
Trustworthiness	• Portability
• Security	• Extensibility
• Integrity	
• Credibility	
Compliance	

04-02-22. IT-Led service quality - Compliance - Definition, desired state, best practices

Module 4: Service concepts, desired states and practices

196

Governance – The effective, ethical management of a company by its executives and managerial levels.

Risk – The ability to effectively and cost-efficiently mitigate risks that can hinder an organization's operations or ability to remain competitive in its market.

Compliance – A company's conformance with regulatory requirements for business operations, data retention and other business practices

Source: https://www.webopedia.com/TERM/G/grc-governance-risk-compliance.html

IT-Led service quality – Compliance – practices

■ <u>Governance, Risk and Compliance (GRC)</u>

IT-Led Services	
Qualities	
Availability	Usability
• Manageability	• Internationalization
• Serviceability	• Accessibility
• Performance	Adaptability
• Reliability	• Interoperability
• Recoverability	• Scalability
• Discoverability	• Elasticity
Trustworthiness	• Portability
• Security	• Extensibility
• Integrity	
• Credibility	
Compliance	

ISACA, with their product, COBIT, is an excellent source of information on GRC.

IT-Led service quality – Usability – definition

Usability

The ability to facilitate ease of operation by users, including users with different needs, e.g., with different native languages, low vision, etc.

Source: Adapted from www.opengroup.org/public/arch/p3/trm/tx/tx_quals.htm

IT-Led Services	
Qualities	
Availability	Usability
• Manageability	• Internationalization
• Serviceability	• Accessibility
• Performance	Adaptability
• Reliability	• Interoperability
• Recoverability	• Scalability
• Discoverability	• Elasticity
Trustworthiness	• Portability
• Security	• Extensibility
• Integrity	
• Credibility	
• Compliance	

Services can run the gamut from "one size fits all", to supporting multiple languages, to supporting people with low or no vision, or hearing, and so on.

IT-Led service quality – Usability – desired state

Performance is effective when...

- All users indicate our services and the service management system are easy to operate
- Are services meet regulatory requirements and the needs of our users, including multi-language capability and the capability to support, for example, users with low or no vision or hearing.

IT-Led Services	
Qualities	
Availability	Usability
• Manageability	• Internationalization
• Serviceability	• Accessibility
• Performance	Adaptability
• Reliability	• Interoperability
• Recoverability	• Scalability
• Discoverability	• Elasticity
Trustworthiness	• Portability
• Security	• Extensibility
• Integrity	
• Credibility	
• Compliance	

See https://www.ada.gov/pcatoolkit/chap5toolkit.htm for some examples of ADA compliant websites.

IT-Led service quality – Usability – practices

- **User Experience (UX)**
- **Usability** (website)
- **Usability** (mobile app)
- **Internationalization and localization**
- **Accessibility** (website)
- **Accessibility** (mobile app)

IT-Led Services	
Qualities	
Availability	Usability
• Manageability	• Internationalization
• Serviceability	• Accessibility
• Performance	Adaptability
• Reliability	• Interoperability
• Recoverability	• Scalability
• Discoverability	• Elasticity
Trustworthiness	• Portability
• Security	• Extensibility
• Integrity	
• Credibility	
• Compliance	

Internationalization and localization includes support for different languages, but also currencies, VAT, and so on.

IT-Led service quality – Internationalization – definition

Internationalization

The ability to support multi-lingual and multi-cultural requirements

Source: Adapted from www.opengroup.org/public/arch/p3/trm/tx/tx_quals.htm

IT-Led Services	
Qualities	
Availability	Usability
• Manageability	Internationalization
• Serviceability	• Accessibility
• Performance	Adaptability
• Reliability	• Interoperability
• Recoverability	• Scalability
• Discoverability	• Elasticity
Trustworthiness	• Portability
• Security	• Extensibility
• Integrity	
• Credibility	
• Compliance	

Here's a good list of multi-cultural considerations - www.webanddesigners.com/multicultural-web-design/. It includes things like data and time formats, time zones, languages, currency, and colors with cultural significance to avoid or use.

IT-Led service quality – Internationalization – desired state

Performance is effective when...

- For the services and service management system components that require it, support for multi-lingual and multi-cultural requirements is available in our services and the SMS, and stakeholders are pleased with its qualities

IT-Led Services	
Qualities	
Availability	Usability
• Manageability	Internationalization
• Serviceability	• Accessibility
• Performance	Adaptability
• Reliability	• Interoperability
• Recoverability	• Scalability
• Discoverability	• Elasticity
Trustworthiness	• Portability
• Security	• Extensibility
• Integrity	
• Credibility	
• Compliance	

Obviously, if you do not have a presence, say, in Canada or the UK, you may not have to be concerned with these issues. But if you plan such growth, better to set your systems up (and choose platforms) that support it, than to have to later retool to do so—it can be very painful.

IT-Led service quality - Internationalization - practices

- **Internationalization and localization** (website)
- **Internationalization and localization** (mobile apps)

IT-Led Services	
Qualities	
Availability	Usability
• Manageability	Internationalization
• Serviceability	• Accessibility
• Performance	Adaptability
• Reliability	• Interoperability
• Recoverability	• Scalability
• Discoverability	• Elasticity
Trustworthiness	• Portability
• Security	• Extensibility
• Integrity	
• Credibility	
• Compliance	

Internationalization is the quality of being easily adaptable for multiple languages, regions and cultures, through things like using Unicode character sets, using text labels that can be translated instead of text embedded in graphics, leaving space in layouts for different languages, avoiding local jargon or colloquialisms in written text, using or avoiding colors, etc. with specific cultural meaning, accommodating different time and date layouts and time zones, as well as different currencies, and so forth.

Localization is the process of adapting a service to a particular language, region or culture.

IT-Led service quality – Accessibility – definition

Accessibility

The ability to support access to services to users with disabilities, such as visual, hearing, physical or cognitive impairments, for example, color blindness, Dyslexia, etc.

Source: Adapted from en.wikipedia.org/wiki/Computer_accessibility

IT-Led Services	
Qualities	
Availability	Usability
• Manageability	• Internationalization
• Serviceability	Accessibility
• Performance	Adaptability
• Reliability	• Interoperability
• Recoverability	• Scalability
• Discoverability	• Elasticity
Trustworthiness	• Portability
• Security	• Extensibility
• Integrity	
• Credibility	
• Compliance	

Access to services for the impaired can be facilitate through specialized hardware and software, but services and their components must have the quality of accessibility to support that access. For example, if a person with low vision has a screen reader device, the service or its components must support use of such device. So for example, PDF files can be read with a screen reader, but if the PDF files don't' have transcripts for videos, or text labels behind graphics and buttons, a low vision person cannot "see" this information—in other words, for this particular scenario, the system lacks the quality of accessibility.

IT-Led service quality – Accessibility – desired state

Performance is effective when...

- For the services and service management system components that require it, support for accessibility is available in our services as required by customer and user needs and applicable standards and regulations, including, where applicable, support for specialized hardware and software for accessibility, and support for access by users with disabilities, such as visual, hearing, physical or cognitive impairments, for example, color blindness or Dyslexia.

IT-Led Services	
Qualities	
Availability	Usability
• Manageability	• Internationalization
• Serviceability	Accessibility
• Performance	Adaptability
• Reliability	• Interoperability
• Recoverability	• Scalability
• Discoverability	• Elasticity
Trustworthiness	• Portability
• Security	• Extensibility
• Integrity	
• Credibility	
• Compliance	

For internationalization and accessibility, as well as for things like security and compliance, there are a lot of considerations. Often, the levels of expertise required in just one of these areas are well beyond what some individuals and businesses have or can afford in-house. That is another reason why SaaS, PaaS and IaaS offerings have grown in popularity—they have the wherewithal to build things like compliance and internationalization into their offerings, so you don't have to keep up on the latest requirements and needs and capabilities, you just need to use the capabilities of the offering.

IT-Led service quality – Accessibility – practices

- ■ <u>ADA compliance</u> (for websites)
- ■ <u>ADA compliance</u> (for mobile apps)
- ■ <u>ADA compliance</u> (for software)
- ■ <u>Specialized HW/SW for accessibility</u>

IT-Led Services	
Qualities	
Availability	Usability
• Manageability	• Internationalization
• Serviceability	▸ Accessibility
• Performance	Adaptability
• Reliability	• Interoperability
• Recoverability	• Scalability
• Discoverability	• Elasticity
Trustworthiness	• Portability
• Security	• Extensibility
• Integrity	
• Credibility	
• Compliance	

There is even a browser extension that helps color blind persons better see the web; see https://www.pcworld.com/article/2919980/this-chrome-extension-helps-color-blind-users-see-the-web.html

IT-Led service quality – Adaptability – definition

Adaptability

The ability to adapt to or be adapted for changes external to the service, including interoperability, scalability, portability (of data, people, applications, and components), extensibility, and the ability to accept new functionality, and the ability to offer access to services in new paradigms, such as ChatOps and microservices architecture.

Source: Adapted from www.opengroup.org/public/arch/p3/trm/tx/tx_quals.htm

IT-Led Services	
Qualities	
Availability	Usability
• Manageability	• Internationalization
• Serviceability	• Accessibility
• Performance	Adaptability
• Reliability	• Interoperability
• Recoverability	• Scalability
• Discoverability	• Elasticity
Trustworthiness	• Portability
• Security	• Extensibility
• Integrity	
• Credibility	
• Compliance	

One key criteria for DevOps tool chains is that they have an API and a command line interface, and not just a UI, so you can string them together into something that adds value, sometimes in high-value, unanticipated ways, as in ChatOps. So, tools that are UI-only could be said to have low adaptability, and those with an API and command line interface, high adaptability.

IT-Led service quality - Adaptability - desired state

Performance is effective when...

- It is not impossible or hard to adapt services or the service management system over time and as circumstances change
- Our services and SMS have the right level of adaptability, including interoperability, scalability, portability, extensibility, and the ability to accept new functionality, and the ability to offer access to services in new paradigms, such as ChatOps and microservices architecture.

IT-Led Services	
Qualities	
Availability	Usability
• Manageability	• Internationalization
• Serviceability	• Accessibility
• Performance	Adaptability
• Reliability	• Interoperability
• Recoverability	• Scalability
• Discoverability	• Elasticity
Trustworthiness	• Portability
• Security	• Extensibility
• Integrity	
• Credibility	
• Compliance	

Adaptability is a matter of degree, and the right balance that is cost-effective. Some platforms and systems are inherently more interoperable, scalable, portable, and extensible than others, and we should look to these qualities when we evaluate what goes into our service components and into our tool chains.

IT-Led service quality – Adaptability – practices

- **API and CL interface, not just UI**
- **ChatOps**

IT-Led Services	
Qualities	
Availability	Usability
• Manageability	• Internationalization
• Serviceability	• Accessibility
• Performance	Adaptability
• Reliability	• Interoperability
• Recoverability	• Scalability
• Discoverability	• Elasticity
Trustworthiness	• Portability
• Security	• Extensibility
• Integrity	
• Credibility	
• Compliance	

04-02-26. Service qualities – Adaptability – Definition, desired state, best practices
Module 4: Service concepts, desired states and practices

209

https://docs.stackstorm.com/chatops/chatops.html defines ChatOps as bringing work that is already happening in the background today into a common chatroom. By doing this, you are unifying the communication about what work should get done with the actual history of the work being done (both manually and through automation, including timestamps). Things like deploying code from Chat, viewing graphs from a TSDB or logging tool, or creating new Jira tickets...all of these are examples of tasks that can be done via ChatOps. Besides enhancing collaboration, ChatOps provides a timestamped record of who did what, when, useful as input to blameless postmortems.

IT-Led service quality - Interoperability - definition

Interoperability

The ability to communicate, exchange data, and work with other services.

IT-Led Services	
Qualities	
Availability	Usability
• Manageability	• Internationalization
• Serviceability	• Accessibility
• Performance	Adaptability
• Reliability	Interoperability
• Recoverability	• Scalability
• Discoverability	• Elasticity
Trustworthiness	• Portability
• Security	• Extensibility
• Integrity	
• Credibility	
• Compliance	

Source: Adapted from https://en.wikipedia.org/wiki/Interoperability
04-02-27. IT-Led service quality - Interoperability - Definition, desired state, best practices

Systems that have the quality of being interoperable can be strung together to add more value. So for example, a CRM system may integrate with a system that provides data on businesses, so that contact information, and other vital statistics for each system are auto populated into CRM.

IT-Led service quality - Interoperability - desired state

Performance is effective when...

- It is possible and easy for all services and SMS components to communicate, exchange data, and work with all other services and SMS components, to the extent that it is useful and cost-effective to have this capability, without any special effort required to do so.

IT-Led Services	
Qualities	
Availability	Usability
• Manageability	• Internationalization
• Serviceability	• Accessibility
• Performance	Adaptability
• Reliability	Interoperability
• Recoverability	• Scalability
• Discoverability	• Elasticity
Trustworthiness	• Portability
• Security	• Extensibility
• Integrity	
• Credibility	
• Compliance	

04-02-27. IT-Led service quality - Interoperability - Definition, desired state, best practices

One key thing about interoperability is that if a system has it as a quality, it requires no special effort to integrate it with another compatible system. So for example, google Chrome extensions just snap right into the browser, boom, you're done. You plug an HDMI cable into an HDMI port, and it just works, no special effort.

Often, interoperability is supported by standards like TCP/IP, HTML, and so on, as well as hardware interface standards.

■ Achieving interoperability

IT-Led Services	
Qualities	
Availability	Usability
• Manageability	• Internationalization
• Serviceability	• Accessibility
• Performance	Adaptability
• Reliability	Interoperability
• Recoverability	• Scalability
• Discoverability	• Elasticity
Trustworthiness	• Portability
• Security	• Extensibility
• Integrity	
• Credibility	
• Compliance	

04-02-27. IT-Led service quality – Interoperability – Definition, desired state, best practices

Module 4: Service concepts, desired states and practices

According to https://en.wikipedia.org/wiki/Interoperability#Achieving_software_interoperability, software interoperability can be achieved in five ways:

1. Product testing

2. Product engineering

3. Industry/community partnership

4. Common technology and IP, and

5. Standard implementation

IT-Led service quality – Scalability – definition

Scalability

The ability of a service and its components to grow or shrink in capacity or performance appropriately in proportion to the demands placed on by the environment in which it operates.

Source: adapted from:

www.opengroup.org/public/arch/p3/trm/tx/tx_quals.htm

IT-Led Services	
Qualities	
Availability	Usability
• Manageability	• Internationalization
• Serviceability	• Accessibility
• Performance	Adaptability
• Reliability	• Interoperability
• Recoverability	Scalability
• Discoverability	• Elasticity
Trustworthiness	• Portability
• Security	• Extensibility
• Integrity	
• Credibility	
• Compliance	

IT-Led service quality - Scalability - desired state

Performance is effective when...

- Our services (and the service management system) can scale up / down and out / in, as needed, at a rate of speed and cost that meets our needs, in proportion to the demands placed on them by the environment within which they operate, and to the extent that the level of scalability is cost-effective.

IT-Led Services	
Qualities	
Availability	Usability
• Manageability	• Internationalization
• Serviceability	• Accessibility
• Performance	Adaptability
• Reliability	• Interoperability
• Recoverability	Scalability
• Discoverability	• Elasticity
Trustworthiness	• Portability
• Security	• Extensibility
• Integrity	
• Credibility	
• Compliance	

Scalability

The ability to grow or shrink in capacity or performance appropriately in proportion to demands of the environment in which it operates

Source: adapted from www.opengroup.org/public/arch/p3/trm/tx/tx_quals.htm

IT-Led service quality – Scalability – practices

- Scale up / down, or scale out / in
- Elasticity and scaling differences (cloud)
- Scaling (cloud)

IT-Led Services	
Qualities	
Availability	Usability
• Manageability	• Internationalization
• Serviceability	• Accessibility
• Performance	Adaptability
• Reliability	• Interoperability
• Recoverability	Scalability
• Discoverability	• Elasticity
Trustworthiness	• Portability
• Security	• Extensibility
• Integrity	
• Credibility	
• Compliance	

Scaling can be vertical (up and down) or horizontal (in or out).

IT-Led service quality – Elasticity – definition

Elasticity

The ability through automation, to scale automatically to adapt to workload changes by provisioning and de-provisioning resources in an autonomic manner, such that at each point in time available resources match current demand as closely as possible.

Source: Adapted from en.wikipedia.org/wiki/Elasticity_(cloud_computing)

IT-Led Services	
Qualities	
Availability	Usability
• Manageability	• Internationalization
• Serviceability	• Accessibility
• Performance	Adaptability
• Reliability	• Interoperability
• Recoverability	• Scalability
• Discoverability	Elasticity
Trustworthiness	• Portability
• Security	• Extensibility
• Integrity	
• Credibility	
• Compliance	

https://www.stratoscale.com/blog/cloud/difference-between-elasticity-and-scalability-in-cloud-computing/ points out that scalability and elasticity are not the same, as scalability can be accomplished by simply over-provisioning resources statically. In elasticity, we scale dynamically, up and down, through automation, so that capacity meets demand.

IT-Led service quality - Elasticity - desired state

Performance is effective when...

- Services (and the SMS) can adapt automatically to workload changes by provisioning and de-provisioning resources in an autonomic manner, such that at each point in time available resources match current demand as closely as possible.

IT-Led Services	
Qualities	
Availability	Usability
• Manageability	• Internationalization
• Serviceability	• Accessibility
• Performance	Adaptability
• Reliability	• Interoperability
• Recoverability	• Scalability
• Discoverability	Elasticity
Trustworthiness	• Portability
• Security	• Extensibility
• Integrity	
• Credibility	
• Compliance	

04-02-29. IT-Led service quality - Elasticity - Definition, desired state, best practices

https://en.wikipedia.org/wiki/Elasticity_(cloud_computing) provides an example of elasticity, as follows: say a service provider wants to run a website on an IaaS cloud. At moment, the website is unpopular and a single machine (most commonly a virtual machine) is sufficient to serve all web users. Suddenly, the website becomes popular, for example, as a result of a flash crowd, and a single machine is no longer sufficient to serve all users. Based on the number of web users simultaneously accessing the website and the resource requirements of the web server, it might be that ten machines are needed. An elastic system should immediately detect this condition and provision nine additional machines from the cloud, so as to serve all web users responsively. Now let's say the website becomes unpopular again. The ten machines that are currently allocated to the website are mostly idle and a single machine would be sufficient to serve the few users who are accessing the website. An elastic system should immediately detect this condition and deprovision nine machines and release them to the cloud.

IT-Led service quality - Elasticity - practices

- <u>Autoscaling</u> AWS
- <u>Autoscaling</u> Azure

IT-Led Services	
Qualities	
Availability	Usability
• Manageability	• Internationalization
• Serviceability	• Accessibility
• Performance	Adaptability
• Reliability	• Interoperability
• Recoverability	• Scalability
• Discoverability	Elasticity
Trustworthiness	• Portability
• Security	• Extensibility
• Integrity	
• Credibility	
• Compliance	

Autoscaling is the process of dynamically allocating resources to match demand to meet performance requirements.

IT-Led service quality – Portability – definition

Portability

The ability for the service to be moved or copied from one environment to another, including data, people, applications, and components.

Source: Adapted from www.opengroup.org/public/arch/p3/trm/tx/tx_quals.htm

IT-Led Services	
Qualities	
Availability	Usability
• Manageability	• Internationalization
• Serviceability	• Accessibility
• Performance	Adaptability
• Reliability	• Interoperability
• Recoverability	• Scalability
• Discoverability	• Elasticity
Trustworthiness	➡ Portability
• Security	• Extensibility
• Integrity	
• Credibility	
• Compliance	

For example, software that is portable could be moved from one machine platform to another, say, a VM that could move from Hyper-V to a VMware environment and vice versa.

IT-Led service quality – Portability – desired state

Performance is effective when...

- Services and their components and the SMS and its components can be moved or copied from one environment to another, including data, people, applications and components, without inordinate time or special effort being required

IT-Led Services	
Qualities	
Availability	Usability
• Manageability	• Internationalization
• Serviceability	• Accessibility
• Performance	Adaptability
• Reliability	• Interoperability
• Recoverability	• Scalability
• Discoverability	• Elasticity
Trustworthiness	➡ Portability
• Security	• Extensibility
• Integrity	
• Credibility	
• Compliance	

Portability can also be due to using a standard format, as when a .JPEG file can be sent to and viewed on say, a Windows or Mac device, or an iOS or Android phone.

IT-Led service quality – Portability – practices

- **Portability (Software)**
- **Portability (Cloud)**
- **Portability** (Mobile Apps)
- **Portability** (VMs and Containers)

IT-Led Services	
Qualities	
Availability	Usability
• Manageability	• Internationalization
• Serviceability	• Accessibility
• Performance	Adaptability
• Reliability	• Interoperability
• Recoverability	• Scalability
• Discoverability	• Elasticity
Trustworthiness	Portability
• Security	• Extensibility
• Integrity	
• Credibility	
• Compliance	

The quality of portability can apply, in a greater or lesser degree, to all elements that make up a service, for software, to cloud resources, to mobile apps, VMs, and containers.

Extensibility

The ability of a service or its components to accept new functionality.

Source:
www.opengroup.org/public/arch/p3/trm/tx/tx_quals.htm

IT-Led Services	
Qualities	
Availability	Usability
• Manageability	• Internationalization
• Serviceability	• Accessibility
• Performance	Adaptability
• Reliability	• Interoperability
• Recoverability	• Scalability
• Discoverability	• Elasticity
Trustworthiness	• Portability
• Security	▶ Extensibility
• Integrity	
• Credibility	
• Compliance	

Extensibility is different from scalability; scalability means a service ahs the quality of being able to accommodate growth (with elasticity, this scaling can be automated); generally when you scale, the unit you scale with is uniform, e.g., you are scaling storage, compute, network, or some other uniform resource.

In contrast, extensibility means you are able to add something new to the system, something that is not "more of the same", so it doesn't have to be growth-related. For example, you might extend a CRM system by adding a component that auto-populates Hoovers information into the records for each customer and prospect.

IT-Led service quality - Extensibility - desired state

Performance is effective when...

- Services and their components (and the SMS and its components) can accept new functionality without inordinate time and special effort being required to do so.

IT-Led Services	
Qualities	
Availability	Usability
• Manageability	• Internationalization
• Serviceability	• Accessibility
• Performance	Adaptability
• Reliability	• Interoperability
• Recoverability	• Scalability
• Discoverability	• Elasticity
Trustworthiness	• Portability
• Security	Extensibility
• Integrity	
• Credibility	
• Compliance	

Some services and their components lend themselves to extension, some do not. Those that do have the quality of extensibility.

IT-Led service quality - Extensibility - practices

- **Extensibility** (GitHub)
- **Extensibility** (XML)

IT-Led Services	
Qualities	
Availability	Usability
• Manageability	• Internationalization
• Serviceability	• Accessibility
• Performance	Adaptability
• Reliability	• Interoperability
• Recoverability	• Scalability
• Discoverability	• Elasticity
Trustworthiness	• Portability
• Security	Extensibility
• Integrity	
• Credibility	
• Compliance	

Some language development packages allow for extensibility, but it's difficult to understand and implement, so this is yet another factor to consider in choosing a development environment.

Module summary

- Service qualities are the "-ilities" that are how a service behaves, what developers refer to as the non-functional requirements of the service (versus service configuration which is what components a service is made up of, and service features which are what a service does)
- Both Human-Led and IT-Led services have qualities
  - OSM's list of Human-Led service quality characteristics are adopted from Valerie Zeithaml and Mary Jo Bitner's SERVQUAL
  - OSM's list of IT-Led qualities are adapted from the TOGAF taxonomy of service qualities
- Each of these service qualities has a desired state that, for relevant qualities for a particular service, must be achieved and maintained continuously over time and through changing circumstances to ensure value is continuous

IT-Led Services

Qualities

Availability
- Manageability
- Serviceability
- Performance
- Reliability
- Recoverability
- Discoverability

Trustworthiness
- Security
- Integrity
- Credibility
- Compliance

Usability
- Internationalization
- Accessibility

Adaptability
- Interoperability
- Scalability
- Elasticity
- Portability
- Extensibility

Human-Led Services

Qualities

- Reliability
- Responsiveness
- Trustworthiness
- Empathy
- Attractiveness

Module 5

Service Management System (SMS) concepts, desired states and practices (300m)

What is a service management system? What does it look like when it is in its desired state? What are some best practices for getting it there, and keeping it there? We'll cover that here in this module.

OSM Foundation Course Agenda

Module 1: Service management and Open Service Management	60m
Module 2: Stakeholder concepts, desired states and practices	60m
Module 3: Value and value flow concepts, desired states and practices	60m
Module 4: Service concepts, desired states and practices	240m
Module 5: SMS concepts, desired states and practices	300m
Module 6: Summary and exam preparation	120m

Module Objectives

The purpose of this unit is to help you describe what a service management system (SMS) is, its components, desired states, and best practices, including:

05-01-01. Service Management System

Lesson 1: SMS – Design & transition
05-02-01. SMS – Design & transition
05-02-02. Strategy & GRC
05-02-03. Learning & improvement
05-02-04. Development
05-02-05. Release & deployment
05-02-06. Change & configuration

Lesson 2: SMS – Promotion
05-03-01. SMS – Promotion

Lesson 3: SMS – Support
05-04-01. Support
05-04-02. Event handling
05-04-03. Incident handling
05-04-04. Request handling
05-04-05. Problem handling
05-04-06. Major incident & disaster handling

Lesson 4: SMS – Delivery
05-05-01. SMS – Delivery
05-05-02. Stakeholder relations
05-05-03. Administration
05-05-04. Provisioning, metering & billing
05-05-05. Budgeting & accounting

The recommended study period for this unit is a minimum of 300 minutes, or 5 hours.

Service management system (SMS)
The mechanism, fully or partially automated, used by a service provider to direct and control services to achieve and sustainably maintain the desired state of providing value to stakeholders through services through changing needs and technology possibilities. The SMS includes mechanisms to support how services are planned, designed, developed, deployed, promoted, delivered, supported, monitored, measured, reviewed, maintained and improved.

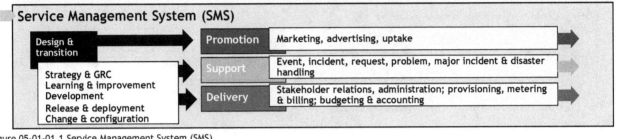

Figure 05-01-01.1 Service Management System (SMS)
05-01-01. Service Management System (SMS) - Definition, desired state, best practices

Module 5: SMS concepts, desired states and practices   229

The service management system or SMS may sound fancy, but it's basically whatever mechanisms you have in place to direct and control services so that they achieve and sustainably maintain the desired state of providing value for stakeholders.

Generally such mechanisms can be said to fall into four groups of related activities:

1. Design & transition – where we figure out what to add, change, or retire—either entire services, features, or feature sets, based on our strategy and what we have learned, especially about changes in stakeholder needs and opportunities and what new technology makes possible.

2. Promotional activities—activities that create uptake of our services, features sets and features.

3. Support – whatever we do to capture and resolve the issues and to fulfill the requests of stakeholders

4. Delivery – the business operations functions of administering services, Provisioning, metering & billing, budgeting for, and accounting for services.

Performance is effective when...

- We continuously and sustainably provide value to stakeholders through services, over time and changing customer and user needs and technology capabilities
- We are achieving and maintaining the desired state for all things worth managing—stakeholders, value flow, services, and the SMS itself
- We continuously improve all aspects of our SMS capabilities, e.g., management, organization, practices, people (skills, knowledge, mindset) policies, objectives, procedures, tools, documents, and resources
- We continuously improve how we to direct and control services, including how services are planned, designed, developed, implemented, deployed, delivered, monitored, measured, reviewed, maintained, and improved

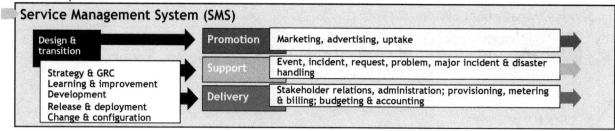

05-01-01. Service Manage System (SMS) - Definition, desired state, best practices

Module 5: SMS concepts, desired states and practices

230

There are some key differences between the SMS as specified in ISO 20000-1 and the guidance contained in OSM, just as there are key difference between ISO's conception of the SMS and traditional ITSM guidance. Some key differences are:

1. ISO emphasizes management's responsibility; OSM positions the outcomes of service management as everyone's job—individuals, teams, and the organization as whole.

2. ISO positions the components of the SMS as processes; OSM positions them as "things worth managing" with desired states to be achieved and maintained; the difference is between a focus on ends (outcomes, or desired states) and means (a process is a potential means).

3. ISO categories by design, delivery, relationship, resolution and control processes; OSM breaks it down into Design & transition, promote, support, and delivery things worth managing with outcomes or desired states.

4. Value flows from the provider to the customer through services as controlled by the SMS in ISO; OSM is similar, except value flows multi-directionally from and to providers and suppliers, customers and users.

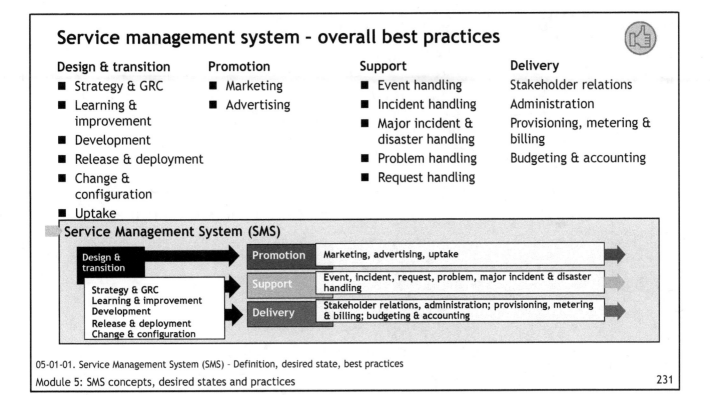

Service management system – overall best practices

Design & transition
- Strategy & GRC
- Learning & improvement
- Development
- Release & deployment
- Change & configuration
- Uptake

Promotion
- Marketing
- Advertising

Support
- Event handling
- Incident handling
- Major incident & disaster handling
- Problem handling
- Request handling

Delivery
Stakeholder relations
Administration
Provisioning, metering & billing
Budgeting & accounting

Service Management System (SMS)

Design & transition →
Strategy & GRC
Learning & improvement
Development
Release & deployment
Change & configuration

Promotion — Marketing, advertising, uptake

Support — Event, incident, request, problem, major incident & disaster handling

Delivery — Stakeholder relations, administration; provisioning, metering & billing; budgeting & accounting

05-01-01. Service Management System (SMS) – Definition, desired state, best practices

As you can see in the figure, a service management system can be broken down into Design & transition, promotion, support, and delivery. OSM handles these a bit differently than traditional ITSM frameworks, which position these as sequential phases and processes.

In OSM, all of these are seen as going on simultaneously, not sequentially. In other words, while of course there is sequence in workflow, it is not helpful to, for example, put incident handling in one "phase" because "that's where it first becomes important".

OSM position these elements not as "processes", but as "things worth managing"—in other words, each represents a key outcome or desired state that we want to achieve and maintain. The focus here is on ends, not means, as we know that a process is just one means to achieve an outcome, and that a broken process (or a focus on process without attention to other enablers), can disable as well as enable sustained achievement of desired outcomes.

Further, in OSM, we seek to focus on outcomes, as opposed to activities, as this is more lightweight, and suited to today's environment and agile practices.

SMS – Design & transition

05-02-01. SMS – Design & transition
05-02-02. Strategy & GRC
05-02-03. Learning & improvement
05-02-04. Development
05-02-05. Release & deployment
05-02-06. Change & configuration

Module 5 Lesson 1

SMS – Design & transition (90m)

In this lesson, we cover the Design & transition group of activities of the SMS.

Design & transition

The service management system (SMS) practice area concerned with Strategy & GRC, Learning & improvement, development, Release & deployment, and Change & configuration of services.

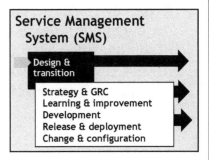

Figure 05-02-01.1 SMS Design & transition
05-02-01. SMS – Design & transition – Definition, desired state, best practices

Module 5: SMS concepts, desired states and practices   233

In this lesson, we cover the Design & transition group of activities of the SMS.

Recall that the purpose of your SMS is to direct and control services so the continuously deliver value to stakeholders, sustainably, over time and through changing circumstances.

This will not happen if you don't have the right strategy for services in the first place, aligned gyroscopically to changes in what stakeholders value and what new technology makes possible, including ensuring they meet any regulatory compliance requirements. This is the strategic learning and improvement loop of the SMS, Strategy & GRC.

Strategy and designs need to be fed through learning loops, and in a dynamic environment, continual improvement is necessary to ensure all things worth managing are as they should be, and if not, are moving gyroscopically towards that desired state—this is the operational and tactical feedback loop of the SMS, Learning & improvement.

The SMS activity group of Development is just what is sounds like—the planning and building of new or changed services, feature sets, and features, qualities (such as availability), and service configuration components, e.g., infrastructure for the service.

Release & deployment is the set of activities within the Design & transition activity group that ensures a quick and quality introduction of new or changed services, features, features sets, qualities, and configuration components. Release & deployment also includes activities for the removal of the same.

Change & configuration is the set of activities that ensures any changes to services are done quickly, with quality, that "what changed?" can be tracked, and that at any time, we have ready visualization everything that constitutes the services we manage.

Performance is effective when...

■ We have the right services, over time and as stakeholder value and technology possibilities change, because we have strategic feedback loops that we use to gyroscopically align our services, features, qualities and configuration to what is valued, over time and through changes in stakeholder needs and technology possibilities

■ We have smooth, unencumbered, automated value flow in our Release & deployment and change & configuration activities, because we all continuously work towards fast time-to-value, with quality and continuous improvement

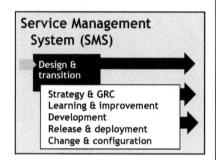

Service Management System (SMS)

Design & transition

Strategy & GRC
Learning & improvement
Development
Release & deployment
Change & configuration

05-02-01. SMS - Design & transition - Definition, desired state, best practices

Recall that the purpose of your SMS is to direct and control services so the continuously deliver value to stakeholders, sustainably, over time and through changing circumstances.

This will not happen if you don't have the right strategy for services in the first place, aligned gyroscopically to changes in what stakeholders value and what new technology makes possible, including ensuring they meet any regulatory compliance requirements. This is the strategic learning and improvement loop of the SMS, Strategy & GRC.

Strategy and designs need to be fed through learning loops, and in a dynamic environment, continual improvement is necessary to ensure all things worth managing are as they should be, and if not, are moving gyroscopically towards that desired state—this is the operational and tactical feedback loop of the SMS, Learning & improvement.

The SMS activity group of Development is just what is sounds like—the planning and building of new or changed services, feature sets, and features, qualities (such as availability), and service configuration components, e.g., infrastructure for the service.

Release & deployment is the set of activities within the Design & transition activity group that ensures a quick and quality introduction of new or changed services, features, features sets, qualities, and configuration components. Release & deployment also includes activities for the removal of the same.

Change and configuration is the set of activities that ensures any changes to services are done quickly, with quality, that "what changed?" can be tracked, and that at any time, we have ready visualization everything that constitutes the services we manage.

SMS – Design & transition – practices

M1

- Business Relationship Management
- Strategy Management
- Service Portfolio Management
- Knowledge Management
- Continual Improvement
- Release and Deployment Management
- Change Management
- (Production) Configuration Management

M2

- The Three Ways (feedback loops)
- Sprint Planning, Sprint Review
- CAMS
- Blameless postmortems
- Governance, Risk and Compliance
- Visible Ops-Style Change Control

- Pets versus Cattle
- Immutable Deployment
- Infrastructure Automation
- Site Reliability Engineering
- Continuous Integration
- Continuous Delivery
- Continuous Deployment
- Blue/green deployment

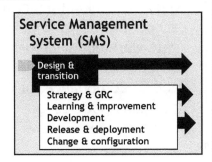

05-02-01. SMS – Design & transition – Definition, desired state, best practices

Module 5: SMS concepts, desired states and practices

235

Traditional ITSM guidance somehow leaves out GRC as a process, and positions BRM, strategy, and portfolio management as processes in a strategy phase.

OSM combines strategy and GRC into one "thing worth managing", because both are "north star" considerations that should inform all planning and decisions and actions in organizations.

SMS – Design & transition – Strategy & GRC – definition

Strategy & GRC

Strategy is the practice of defining key objective and a direction, allocating resources to pursuing that direction, and putting mechanisms in place to guide the achievement of key objectives. GRC is made up of 1) Governance, which describes the leadership, decision-making structure, processes, and accountability that determine how an organization gets work done, 2) Risk, which is helping to achieve objectives by managing risk through internal controls, and 3) Compliance, or ensuring conformance with company policies, governmental regulations, and industry-specific laws.

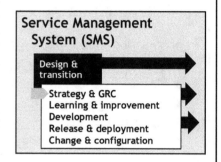

Strategy cannot be decided in a vacuum, it must be informed by constraints; chief among these are governance, risk and compliance considerations.

Performance is effective when...

- We have the right strategy at any given time, which is adjusted continually as stakeholder needs and technology possibilities change, and that strategy is in action in our organization; the strategy includes a strategy for stakeholders: how to keep them in a desired state; for services: which services to offer, and to whom, and which to add, change, or decommission, as well as a strategy for the SMS: for what we need to add, change, or decommission to properly direct and control services

- Our stakeholders, services, and service management system are in compliance with all applicable regulatory and legislative requirements, as well as internal standards and policies; we have the right set of effective policies, controls and required documentation in place to maintain a state of compliance, including regulatory compliance, standards, and security policies

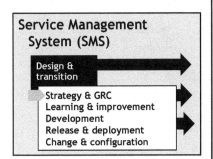

05-02-03. SMS – Design & transition - Learning & improvement - Definition, desired state, best practices

Module 5: SMS concepts, desired states and practices

237

Strategy & GRC

Strategy is the practice of defining key objective and a direction, allocating resources to pursuing that direction, and putting mechanisms in place to guide the achievement of key objectives. GRC is made up of 1) Governance, which describes the leadership, decision-making structure, processes, and accountability that determine how an organization gets work done, 2) Risk, which is helping to achieve objectives by managing risk through internal controls, and 3) Compliance, or ensuring conformance with company policies, governmental regulations, and industry-specific laws.

M1
- Business Relationship Management
- Strategy Management
- Service Portfolio Management

M2
- The Three Ways (feedback loops)
- Sprint Planning, Sprint Review
- CAMS
- Governance, Risk and Compliance

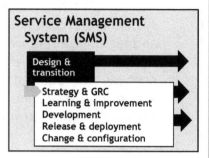

Service Management System (SMS)

Design & transition

Strategy & GRC
Learning & improvement
Development
Release & deployment
Change & configuration

05-02-03. SMS – Design & transition – Learning & improvement – Definition, desired state, best practices

With the M2 SMS, you'll need to put in place more, faster, and tighter feedback loops between BRM, strategy, knowledge management, continual improvement and sprint planning and reviews than in traditional ITSM. You should expect shorter roadmaps, and for the typical unit of work to shift from entire services to feature sets and individual features, and for the whole thing to happen much faster. A core part of strategy will also have to be figuring out what you will do to ensure uptake of new features and feature sets, so that value is realized by consumers; as you ratchet up the pace of deployment, it becomes more difficult for consumers to take up what you are putting down—you'll need a strategy to ensure they do.

You'll also need to integrate GRC in with strategy, which was missing in some traditional ITSM models. As for change control, for Visible Ops-style is recommended. There is a shift-left in configuration management, from managing the logical picture of your configuration, to managing the scripts and templates that create it. Immutable deployment and blue/green deployment have a profound effect on release and change, as do continuous integration, deployment, and delivery.

38   There are no "05-02-02. SMS - Design & transition - Strategy & governance - Definition, desired state, best practices" slides
      Jason Scarpello iMac27, 11/24/2017

Learning & improvement

The set of practice for continual learning and improvement of stakeholders (skills, knowledge, mindset), services, and the SMS, with the overall objective of faster time to value, with quality.

Measuring, analyzing and sharing both quantitative and qualitative data and information about changing business circumstances and new technologies, and historical and projected information on stakeholders, services, and the service management system, and applying the knowledge gained to identifying opportunities for improvement

Making choices for improvements based on value, cost, risk, effort, etc. and assigning priorities to what improvements and enhancements to make to all aspects of stakeholder capabilities (skills, knowledge, mindset), services, and the service management system.

Improvements and enhancements can be anything from individual actions (e.g., something that takes an hour to dispatch) to an improvement project (e.g., something that takes two weeks, a few people, and some money to do). Ensuring legal and regulatory requirements and internal standards are implemented and corresponding policies are followed.

Service Management System (SMS)

Design & transition

Strategy & GRC
Learning & improvement
Development
Release & deployment
Change & configuration

05-02-03. SMS – Design & transition – Learning & improvement – Definition, desired state, best practices

Learning & improvement is the set of practices roughly analogous to traditional ITSM's service strategy and continual service improvement phases; OSM combines strategy, compliance and improvement into one set of practices for the following reasons:

- Both consider the same topic: give our situation, constraints (including regulatory) and scarce resources, where should we invest?

- As the cadence for introducing new services and SMS components and changes and improvements to them increases, as it tends to in more modern IT environments, the utility of separating out this work into two phases decreases, and in fact, the separation may become a blocker

OSM concentrates the Learning & improvement mechanisms that traditional ITSM separates out by process into the Learning & improvement component of the service management system.

Performance is effective when...

- We are continually learning about and improving aspects of stakeholders, services (both IT-Led and Human-Led, including their configuration, functionality and qualities), and the SMS, taking changing stakeholder needs and new technology possibilities into account
- Stakeholders can freely share and access knowledge when it is needed; Because of this, we make better quality decisions and can move faster with less risk
- We review all services, at least annually, after major changes, and three months after the introduction of a new service, and reflect the health of the service and improvement actions back to stakeholders
- We conduct blameless post mortems after disasters, major incidents, problems, and filed changes and releases, for learning, action and improvement
- We regularly review supplier performance against commitments and alignment of supplier goods and services to the support of stakeholders, services and the SMS

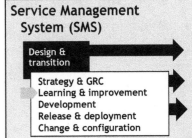

Other indicators of effective performance include:

- We work to understanding of patterns of business activity and forecast demand and build those understandings into the service improvements we make

- We have built effective and efficient mechanisms and feedback loops into our practices, services, and service management system that result in the right actions being taken at the right pace and in the right order to ensure we have the right mix of services, by helping us decide which services and SMS components to add, change, or remove, and that those services and the SMS components are of higher value than alternatives to stakeholders as business circumstances change and new technologies create new possibilities

- We have a continuously evolving vision for what we want to be as a provider, and that vision aligns with stakeholder values, and our portfolio of services aligns with that vision, and where it does not, we have a strategy and action to close the gap by adding, changing, or retiring services and service management system components

- Improvements to services and the SMS are continual, and at the right pace and priority, given available resources

- We have the right set of services in the first place, and where that is not the case, we are working quickly to close the gap by adding, modifying, or removing services, features, quality, and pricing models

- We are continuously aware of how our services stack up against alternatives in terms of price, features, and quality, and where we are not, we move quickly to close the gap; based on this knowledge, we are taking action to close the gap in differentiation by adding, modifying, or removing services, features, quality, and pricing models

M1
- Knowledge Management
- Continual Improvement
- Plan-Do-Check-Act (PDCA)
- CSFs and KPIs
- Service reviews
- Post-implementation reviews
- Trend analysis

M2
- The Three Ways (feedback loops)
- Sprint Planning, Sprint Review
- CAMS
- Blameless postmortems

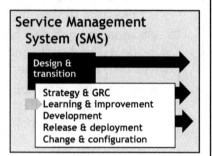

The Deming cycle, is also known as the PDCA or plan-do-check-act cycle.

Conduct blameless postmortems, e.g., after a major incident, problem, disaster, or failed change or release.

Development

Planning, building and testing new and changed stakeholder satisfying mechanisms, value flow mechanisms, services, feature sets, features, qualities, and configuration components, and service management system mechanisms.

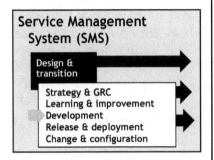

Service Management System (SMS)

Design & transition

Strategy & GRC
Learning & improvement
Development
Release & deployment
Change & configuration

So you can see from this definition that development's scope in OSM is not just applications, but stakeholder mechanisms, services (including their configuration, functionality, and qualities), and the SMS.

Performance is effective when...

- We develop stakeholder mechanisms, and the SMS, and not just services; when we develop services, we get the configuration, functionality, and qualities right
- We continuously listen and improve what and how we Design & transition, with more, faster, and tighter feedback loops between stakeholders, demand and capacity and financial data, strategy & GRC, knowledge management, continual improvement and sprint planning and reviews, shorter roadmaps, and smaller work packages
- We automate and provide self-service for development resources and activities well, for faster time-to-value with quality from our build pipeline
- Both services and the service management system are developed to sustainably meet current and projected stakeholder needs, including the right initial feature set and quality, and an effective mechanism to learn when stakeholder needs change, and vet and introduce new and changed features and quality accordingly
- Designs are consistent, with no unnecessary variations or dependencies

Service Management System (SMS)

Design & transition

Strategy & GRC
Learning & improvement
Development
Release & deployment
Change & configuration

05-02-04. SMS - Design & transition - Development - Definition, desired state, best practices

There are of course many aspects to development, more than can be incorporated within this space and within the scope of a service management foundation course. Having said this, it's important to note what we are developing—not just services, but stakeholder mechanisms and the SMS; that we have a "smaller + faster = better" approach, with a focus on automation and self-service in the build pipeline.

M1
- Design Coordination
- Service Catalog Management
- Service Level Management
- Capacity Management
- Availability Management
- IT Service Continuity Management
- Information Security Management
- Supplier Management

M2
- The Three Ways (feedback loops)
- Sprint Planning, Sprint Review
- CAMS
- Blameless postmortems

- Infrastructure Automation
- Site Reliability Engineering
- Continuous Integration
- Continuous Delivery
- Continuous Deployment
- Lean and Agile Practices
- Application Lifecycle Management
- Test-Driven Development
- User Stories

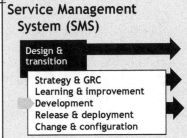

Service Management System (SMS)

Design & transition

Strategy & GRC
Learning & improvement
Development
Release & deployment
Change & configuration

05-02-04. SMS - Design & transition – Development – Definition, desired state, best practices

Module 5: SMS concepts, desired states and practices

244

Some definitions:

Continuous Delivery

Automated implementation of the application build, deploy, test and release processes meant to ensure performance (fast delivery) and conformance (to requirements), enabling users to start using new functionality and qualities quickly to realize value and provide feedback sooner.

Application Lifecycle Management (ALM)

The supervision of, and documentation and tracking of changes to a software application from its initial planning through removal. Source: searchsoftwarequality.techtarget.com/definition/application-lifecycle-management-ALM

SMS – Design & transition – Release & deployment – definition

Release & deployment

Planning, scheduling and controlling the build, test and deployment of releases delivering new services or service management system components, or new or changed features and quality to existing services and SMS components, delivering new functionality and performance quickly with quality, without disrupting existing stakeholders, services, and service management system components. Includes the initial plan, build, test, and deployment of a service or service management system component, as well as major change, such as the transfer of service or service management system component between providers, and removal or services or service management system components. Includes ensuring the release itself is as it should be, the deployment plan and mechanism are good, and that stakeholders, other services, and the SMS are prepared to accept the release (e.g., through training an knowledge transfer); includes ensuring we can back-out of a release.

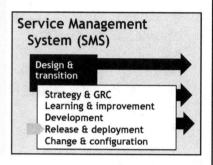

05-02-05. SMS – Design & transition - Release & deployment – Definition, desired state, best practices

A release is a set of related components and changes that we manage as a set, because it makes sense to do so; for example, we could roll out SAP as 5700 changes, one at a time, but it makes more sense to manage it as a release—to put a team on it, to make sure the live environment is ready for the releases, to make sure the release itself is good, and the plan and mechanism to deploy it and back it out if necessary are good, and so on. Each individual change associated with a release still needs to be managed through change control, it's just that it makes sense to also manage at the "zoom level" of a release.

Performance is effective when...

- Stakeholders realize value quickly, with quality, from new or changed services, feature sets, features, and qualities, which are released quickly and with quality, through a pipeline whose flow we automate and continuously improve
- We've put mechanisms in place that make it easy to try new services, feature sets, and features, and roll them back where necessary and useful
- We do a good job at making sure we have the right resources and enough of them when we release and deploy new or changed services, especially when there are concurrent releases and deployments, and resolving schedule and resource conflicts when they do arise
- We have consistency across releases and deployments, no irrational variation
- Cost, schedule, and resource estimates are sufficiently accurate
- We have good feedback loops from stakeholders back to Release & deployment

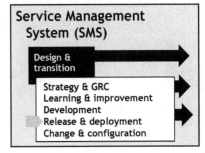

05-02-05. SMS – Design & transition – Release & deployment – Definition, desired state, best practices

With release and deployment management, we are working to get customers and users the services, feature sets, and features and qualities they needs, quickly, with quality, and with continuous improvement, both of services and the build pipeline.

M1
- Release and deployment
- Release policy
- Definitive Media Library

M2
- Continuous Delivery
- Blue/green deployment
- CAMS
- Immutable deployment

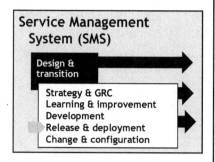

Service Management System (SMS)

Design & transition

Strategy & GRC
Learning & improvement
Development
Release & deployment
Change & configuration

05-02-05. SMS – Design & transition - Release & deployment - Definition, desired state, best practices

A definition:

Continuous Delivery

Automated implementation of the application build, deploy, test and release processes meant to ensure performance (fast delivery) and conformance (to requirements), enabling users to start using new functionality and qualities quickly to realize value and provide feedback sooner.

Change & configuration

Coordinating changes (additions, modifications, deletions) related to stakeholders, services, or the service management system within timelines set by service level targets (SLTs) for different priority changes. Minimizing disruption to stakeholders, services, and the service management system due to the change, and being able to track, "what changed?" precisely enough. Conducting changes efficiently and without much back-out or re-work. Ensuring stakeholders realize value from changes quickly with quality.

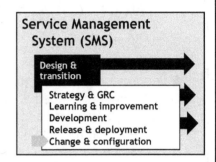

05-02-05. SMS - Design & transition - Release & deployment - Definition, desired state, best practices

Module 5: SMS concepts, desired states and practices

248

Change and configuration are tied at the hip; change is make the actual changes well, and configuration is making sure that we have a good logical picture of what we have out there, and for each component of our configuration, know what it looks like, to a level of detail that is useful to us, and how it relates to other things we have out there.

Performance is effective when...

- We get stakeholders the changes they need quickly and with quality, so they can start realizing value as soon as possible.

- We minimize disruptions due to changes, and are able to track, "what changed?" precisely enough; and we don't have a lot of backed out changes, but when we must, we can back out easily.

- The reality and how stakeholders see it, is how we handle changes as not as slowing down needed changes, but as helping those making changes move quickly, taking appropriate risks and facilitating speed with quality

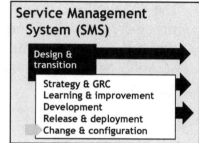

- We ensure a logical model of IT Services, Assets and IT Components needed is defined, controlled, maintained and kept accurate and detailed enough as a source of information for fact-based management of IT services and to comply with corporate governance requirements.

05-02-06. SMS - Design & transition - Change & configuration - Definition, desired states and practices

Module 5: SMS concepts, desired states and practices

Some other indicators of effective performance include:

We minimize the number of eyeballs that need to be on changes, and work relentlessly to root out unnecessary complexity, variation, and dependencies, so that fewer stakeholders must be involved in looking at few changes, and can focus on the ones that really need their attention

We know, precisely enough, what configuration items we have, how they relate to one another and to relevant classes or groups, and what their configuration is to a sufficient level of detail and accuracy to do the work that we do well, and all stakeholders who need the information have easy access to this information when they need it, e.g., when troubleshooting or planning; this includes configuration items that are internal to us as the provider, and external to us, where they are components of our stakeholders, services, and the SMS; this goes for production and pre-production configuration items

We build out configuration items that make up components of stakeholders, services, and the SMS from known good sources and keep them in a known good state.

We have a baseline of configuration items as a reference point and all the snapshots we need for reference and to be able to restore to a prior state, or for other purpose we have

We can tell multiple versions of the same configuration items apart easily, and can easy compare and see how they are the same and different

We can produce accurate information on stakeholder, service, and SMS components and their configuration, versions, and relationship among components easily, e.g., for audits.

Our configuration is under control—it is always the case that the components, versions, and relations between components of stakeholders, services and the SMS are in a known and controlled state, with demonstrable integrity. And we have visualization into current state, planned state, and past state for all configuration items when we need it

SMS – Design & transition – Change & configuration – practices

M1

- Change management
- Configuration management
- Baselining and snapshotting
- CMDB

M2

- Microservices architecture / REST
- Variation reduction
- Immutable deployment
- Infrastructure-as-code

- Service Discovery
- Cloud
- Standardizing resource configurations
- Software configuration management

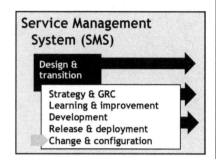

The general shift from traditional ITSM practices is due to a shift from traditional to cloud/mobile environments. For cloud/mobile, suitable practices are needed to fit the target environment to be managed.

Lesson summary

A Service management system (SMS) is a set of specialized organizational capabilities (management, organization, processes, knowledge, people (experience, skills, relationships)) and interrelated or interacting elements (including policies, objectives, procedures, tools, documents, and resources) service providers use to direct and control services, including how services are planned, designed, developed, implemented, deployed, delivered, monitored, measured, reviewed, maintained, and improved. The aim of the SMS is to achieve and sustainably maintain the desired state of providing value to customers and users in the form of services.

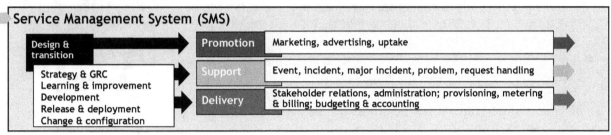

Lesson summary

- Design & transition are the set of SMS practice areas concerned with:
  - Strategy & GRC — ensuring there is a strategy and governance and compliance mechanisms, and that services are continually aligned to it
  - Learning & improvement of services—ensuring the set of services is continuously aligned to stakeholder needs over time and through changing circumstances; this includes knowledge management, and continuous improvement
  - Development of services—the initial planning and building of services
  - Release & deployment — the initial release and deployment of services
  - Change & configuration of services — changes to services and subsequent releases and deployments, through the lifecycle of the service, including removal
- Each of these practice areas has a desired state that must be achieved and maintained continuously over time and through changing circumstances to ensure value continually flows to all stakeholders; these desired states can be achieved and maintained through best practices

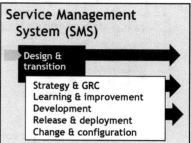

Service Management System (SMS)

Design & transition

Strategy & GRC
Learning & improvement
Development
Release & deployment
Change & configuration

Promotion

05-03-01. Service Management System – Promotion – Definition, desired state, best practices

Module 5 Lesson 2

SMS concepts, desired states and practices – Promotion (30m)

The purpose of this unit is to help you describe what a service management system is, its typical components, their desired states, and best practices for achieving them, including the following SMS components in the category Promotion.

Promotion

Promotion consist of lead generation, enrolling subscribers, and driving uptake, where:

- Lead generation (a strong call to action and communication of the benefits of a service) for a service; for external customers, this is called advertising

- Enrolling subscribers (finding out what people want, why, and for how much in the case of external customers or internal chargeback, or getting them to initially use the service in the case of showback) to services. For external customers, this is sales, converting an inquiry or lead into a subscription.

- Driving uptake (of the use of new services, feature sets and features as they roll out) to enhance value for all stakeholders, and to avoid, "shelfware", i.e., to ensure they are used and value is fully realized by stakeholders.

Figure 05-03-01.1 SMS Promotion   Source: adapted from www.ama.org/aboutama/pages/definition-of-marketing.aspx
05-03-01. SMS – Promotion – Definition, desired state, best practices

Module 5: SMS Definitions, Desired States and Practices   254

Both internal and external services need to be taken up in order for value to be realized. Externally, to external customers, this is marketing, advertising, and sales. For internal customers, it's internal marketing and "selling". Once services are subscribed to, either internally or externally, it is vital that they get used (the pejorative term for software that is paid for but not used is "shelfware") in order for value to be realized fully by stakeholders.

SMS – Promotion – Desired state

Performance is effective when...

- New customers are buying our services, and referring others to buy our services in numbers meeting or exceeding our revenue and profit requirements
- Existing customers are staying with us and referring others to buy our services
- New users are using our services, including existing, new and changing services and new and changing service functionality and qualities, and referring others to use them in numbers meeting or exceeding our requirements
- Existing users are staying with us and referring others to use our services

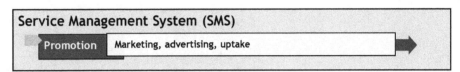

Service Management System (SMS)

Promotion — Marketing, advertising, uptake

SMS – Promotion – practices

M1
- <u>Service catalog</u>

M2
- <u>Lead generation</u> and <u>Enrollment</u>
- <u>Driving uptake</u>

Service Management System (SMS)

| Promotion | Marketing, advertising, uptake |

05-03-01. SMS – Promotion – Definition, desired state, best practices
Module 5: SMS Definitions, Desired States and Practices

256

Some best practices for Human-Led services promotion:

Delivery is a services firm's most valuable sales tool. *("The PSF 50", pg. 156, "How to Buy/Sell Professional Services")* A services firm's ability to sell begins and ends with the ability of its consultants to deliver. A sale does not end when an engagement begins—the only thing a client has at the beginning of an engagement is the *promise* of results. A sale ends when all results have been *achieved* to the client's satisfaction.

Existing clients are our highest probability prospects. *("Managing the Professional Service Firm", pg. 97)* Existing clients are a services firm's best source of new business. When dealing with existing clients, services firms have already earned their trust and confidence, they often understand other aspects of the sales process important to the final decision, and there is often little or no competition, especially in cases where the firm discovers the need.

The best way to sell is to care. *("Managing the Professional Service Firm", pp. 71, 120)* The essential nature of services revolves around relationships, not technical matters. Clients shop for trust, confidence, and peace of mind as much as they do technical expertise. They respond most favorably to consultants with an interest in their problems and a sincere desire to help. Therefore, the best means of attracting clients is to actually care about helping clients succeed.

The marketing of services must be a seduction, not an assault. *("Managing the Professional Service Firm", pg. 122)* Marketing services is most effective when it demonstrates competence, not when it asserts it. Successful marketing of services is ultimately about attracting clients in a manner that *they* want to take the next step in the relationship.

The goal of marketing is to create face-to-face interaction. *("Managing the Professional Service Firm", pg. 121)* The ultimate goal of all marketing efforts in a services firm is to generate face-to-face dialogues with prospects. Individualized content delivered through face-to-face interaction is much more effective than general messages broadcast to large audiences.

Lesson summary

- Services have to be purchased and used for all stakeholders (customers, users, the provider, and supplier) to realize value
- Value is a function of uptake of services, including existing, new, and changing functionality and service qualities; the goal of the provider is to have the right services and functionality and service qualities in the first place, and as time passes and customer and user circumstances and technology possibilities change, to add, modify, and remove services, functionality, and qualities to ensure services continuously provide value; but this is not enough; providers must ensure uptake of services, including existing and new and changing functionality and service qualities, to ensure customers and users fully realize value
- The promotion practice area of the service management system is aimed at ensuring uptake of services through marketing, advertising, and sales of external services to external customers and users, or the equivalent activities for internal services for internal customers and users
- The promotion practice area has a desired state that must be achieved and maintained continuously over time and through changing circumstances to ensure value continually flows to all stakeholders; this desired states can be achieved and maintained through best practices

Service Management System (SMS)	
Promotion	Marketing, advertising, uptake

Module 5 Lesson 3

SMS concepts, desired states and practices - Support (90m)

The purpose of this unit is to help you list and describe the what a service management system is, its typical components, their desired states, and best practices for achieving them, including the following SMS components in the category Support.

Support

The component of the SMS that directs, controls and executes the day-to-day support of services, including handling events, requests, incidents, problems, major incidents and disasters. Contrast support with delivery, which is the component of the SMS that directs, controls and executes the day-to-day operation of services, including stakeholder relations, administration, provisioning, metering, and billing, and budgeting and accounting. Support is characterized by handling of customer and user requests within a timeline set in an SLA, as well as major incidents, problems, and disasters; while there are reactive aspects of the delivery components of the SMS, they are primarily characterized by planned activities on a set schedule, whereas support is primarily characterized by responding to customer and user requests, as well as problems, major incidents and disasters, with proactive aspects being important, but less frequent, than those that are unplanned.

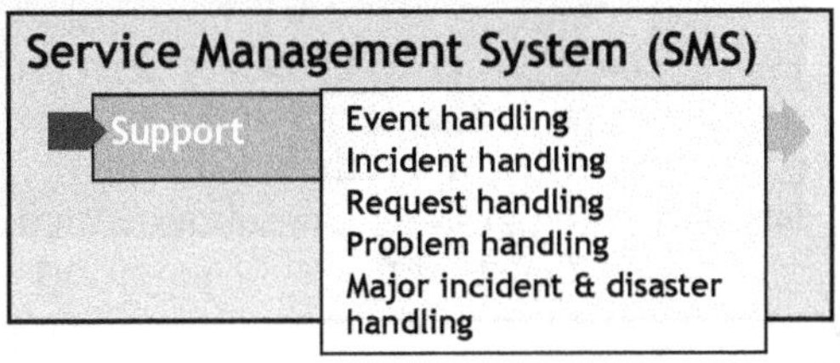

Service Management System (SMS)

Support

Event handling
Incident handling
Request handling
Problem handling
Major incident & disaster handling

Figure 05-04-01.1 SMS Support
05-04-01. SMS - Support - Definition, desired state, best practices

Module 5: Service Management Systems concepts, desired states and practices

259

We want to detect events related to services must be detected and notifications sent out, so that desired control actions (either automated or manual) can be taken, including triggering incident handling for exceptions we have trapped. We also want to make sure all transactions that come in from our users, either through the portal or some other channel, get handled—both requests and incidents, either through automation, or by human intervention, or by some combination of the two. Lastly, we want to make sure major incidents (which lie somewhere between a regular incident and an all-out disaster, and which, if left alone, tend to become disasters) and disasters get handled properly. Handling in all these cases includes both proactive provisions for these transactions and scenarios, as well as the reactive dispatching of them when they do occur.

SMS - Support - desired state

Performance is effective when...

- Stakeholders typically do not require support, because our services just work and are easy to grasp and use; in the rare occasion that they do require support, stakeholders are satisfied with the level of support they receive, both with the result they get and with the interaction / experience of support
- Stakeholders indicate that we detect all events (informational, warning, or exception) related to services that are worth knowing about and managing, and send out notifications so that desired control actions (either automated or manual) can be taken, including triggering incident handling for exceptions
- Stakeholders indicate that all requests and incidents that come in from users, either through the portal or another channel, get handled, through automation, human intervention, or some combination of the two.
- Stakeholders agree that we make sure major incidents (which lie somewhere between a regular incident and an all-out disaster, and which, if left alone, tend to become disasters) and disasters get handled properly, including both proactive provisions for these transactions and scenarios, and reactive dispatching when they do occur
- We automate and continuously improve to reduce support cost, timescales and effort

Service Management System (SMS)

Support
- Event handling
- Incident handling
- Request handling
- Problem handling
- Major incident & disaster handling

05-04-01. SMS - Support - Definition, desired state, best practices

In the end, we want to provide responsive support at the right cost.

SMS - Support - practices

M1

- <u>Event handling</u>
- <u>Incident handling</u>
- <u>Major incident & disaster handling</u>
- <u>Disaster handling</u>
- <u>Request handling</u>
- <u>Problem handling</u>
- <u>Self-service portal</u>

M2

- Event handling

- <u>Monitoring and logging</u>
- <u>Incident handling</u>
  - <u>Multi-channel service</u>
- Major incident & disaster handling
  - <u>Incident Command System (ICS)</u>
  - <u>Infrastructure Automation</u>
- Request handling
  - <u>CAMS</u> (for automation)
- <u>Problem handling</u>

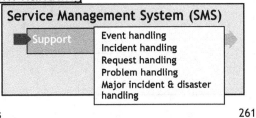

05-04-01. SMS - Support - Definition, desired state, best practices

Module 5: Service Management Systems concepts, desired states and practices

261

Traditional ITSM guidance positions the handling events, incident, major incidents, problems and requests as processes in an operations phase of a lifecycle. OSM positions them as "things worth managing" with desired states to be achieved and maintained. The difference in positioning is the difference in a means (process) versus end (outcome or desired state) model. Traditional ITSM guidance positions disaster handling in a service design phase. OSM places it inline with other support and response processes here; note that the service quality of recoverability associated with disaster preparedness is covered in the service qualities portion of the OSM model.

One example here of a mode 2 practice: Infrastructure automation is based on the cloud-based reality that infrastructure components can and should be treated like code. System specs should be checked into source control, go through a code review whether a build, an automated test. Then you can automatically create uniform instances from the spec and to manage them programmatically. There are so many ITIL challenges you can overcome by infrastructure automation. For example, we can start by applying the DevOps method of infrastructure automation to your ITIL-driven IT Service Continuity, or disaster recovery process. We used to talk a lot about design for availability and a DR site and hardware in another city, keeping two data centers in sync; now because I can script my infrastructure deployment I can just grab my script, point it at the other datacenter, run it and bam, I've got another instance of my infrastructure. Some organizations I work with have gone beyond infrastructure as code, to operations a code (automating every ITIL process with code); alerts, incidents, problems, provisioning, capacity, demand, availability, performance, everything can be reduced to automation through code.

SMS - Support - Event handling - definition

Event handling

Monitoring for and detecting events (changes in state of significance for stakeholders, services, or service management system components, including informational, warnings, and exceptions); raising alerts (notifications), reports, and escalations. In the case of informational and warning event, possibly triggering automated or manual control actions; In the case of exceptions, triggering incident handling.

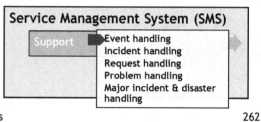

Service Management System (SMS)

Support → Event handling
Incident handling
Request handling
Problem handling
Major incident & disaster handling

SMS – Support – Event handling – desired state

Performance is effective when...

- We have the right set of event detection and alert notification mechanisms set up, so that we can detect all changes of state in stakeholders, services, and the SMS that are significant and may require a control response; These include normal situations, e.g., informational "chron job just ended", to warning "you set a threshold of 15% utilization on this storage array and told me to tell you when it happened; I am telling you"; to exception / abnormal situations "server down"; we have these set up to monitor for all quality of all services, including financials, service levels, continuity, availability, throughput, configuration, security, and compliance, as well as the components of services, including environment (E.g., temperature, fire, water detection), hardware, system software, networks, applications, and DBMS

Service Management System (SMS)

Support	Event handling
	Incident handling
	Request handling
	Problem handling
	Major incident & disaster handling

Event management in OSM is not just about capturing infrastructure or application events, but capturing and notifying about all "things worth managing".

M1

- <u>Event handling</u>

M2

- <u>Instrumentation / Telemetry</u>
- <u>Monitoring</u>
- <u>Cloud Monitoring / Telemetry</u>
- <u>Logging / Instrumentation</u>
- <u>Logging</u> cheat sheet
- <u>API and CL interface, not just UI</u>
- <u>ChatOps</u>

Service Management System (SMS)

Support	Event handling
	Incident handling
	Request handling
	Problem handling
	Major incident & disaster handling

The general trend is to take point solutions with an API and a command line interface, and string them together to create a toolchain, rather than seeking monolithic solutions. Part of the interesting shift in recent year has been that tools that were built assuming a physical on-prem environment haven't necessarily made it over to the cloud, because their core model is so different, and cloud solutions are being produced more through iteration (leading initially to smaller, more point solution offerings), which seems to have lead to this trend.

Incident handling

Fixing incidents (that is, requests to fix something that is broken) related to stakeholders, services, or the service management system within timelines set by service level targets (SLTs) for different priority incidents.

Incident

unplanned interruption to a service, a reduction in the quality of a service or an event that has not yet impacted the service to the customer. Source: ISO/IEC 20000-1:2011(E) 3 Terms and definitions

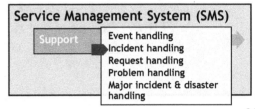

With an incident, something is broken. It could be an error trapped in the infrastructure, or a user who is "broken", i.e., cannot work.

SMS - Support - Incident handling - desired state

Performance is effective when...

- Incidents are handled within service level targets, and stakeholders are satisfied with the professionalism and speed with which we handle incidents
- Incidents are handled effectively and efficiently, while minimizing disruption on stakeholders, services, and the SMS
- Feedback to Learning & improvement for improving both services and the service management system to better handle similar situations going forward
- We employ automation in the handling of incidents
- Incidents are resolved within agree service level targets, including incidents concerning quality of a service, including financials, service levels, continuity, availability, throughput, configuration, security, compliance; in addition to resolving incidents, we analyze cycle time components to see how we can crash MTTR/MTTRS, by taking time out of the cycle, e.g., time to record, respond, resolve, repair/replace, recover

Service Management System (SMS)	
Support	Event handling
	Incident handling
	Request handling
	Problem handling
	Major incident & disaster handling

We work to prevent incidents in the first place, and should they occur, we work to resolve incidents as quickly as possible, certainly within service level targets for MTTR / MTTRS; further, we work to proactively crash the cycle time for incidents through analysis, action, and automation by analyzing where time is spent in the cycle, e.g., time to record, respond, resolve, repair / replace, recover

M1

- Incident management
- Service desk

M2

- Public status pages
- Multi-channel support
- ChatOps
- Devs on Call

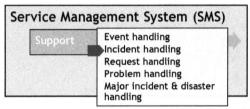

Service Management System (SMS)

Support

Event handling
Incident handling
Request handling
Problem handling
Major incident & disaster handling

05-04-03. SMS - Support - Incident handling - Definition, desired state, best practices

Module 5: Service Management Systems concepts, desired states and practices

267

The DevOps practice Putting Devs On Call for the services they create is based on the golden idea of, "you dealt it, you deal with it"—the idea that Devs will take more care of the quality of what they produce if the will have to suffer the pain of dealing with it later, and not just chucking it over the fence for somebody else to test. The Devs on Call DevOps practice has the greatest impact on traditional ITSM-driven Application Management, Service Desk, and IT Operations functions.

You put Devs on call for the service they create to create a fast feedback loop that helps rapidly improve logging and deployment, and more quickly resolve application issues.

Request handling

Fulfilling requests (for goods or services a stakeholder is entitled to because of their subscription to a service) related to stakeholders, services, or the service management system within timelines set by service level targets (SLTs) for different priority requests. Unlike incidents, nothing is broken—the requestor is just asking for something they are entitled to (in the case that they are not entitled to what they are requesting, these requests are denied or escalated).

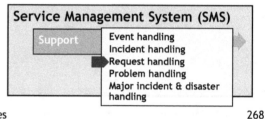

05-04-04. SMS - Support - Request handling – Definition, desired state, best practices

Module 5: Service Management Systems concepts, desired states and practices

268

Note here that requests can be self-service. They can be requested and fulfilled totally manually, or totally automated, on both sides of the equation. Also, note that requestors in this model are not just customers and users. For example, self-service and requests are part of the normal fabric of working within a build pipeline, by provider staff and suppliers.

SMS – Support – Request handling – desired state

Performance is effective when...

- Requests are handled within service level objectives or targets, and stakeholders are satisfied with the professionalism and speed with which their request are handled
- Feedback to Learning & improvement for improving both services and the service management system to better handle similar situations going forward
- We support all request channels stakeholders are inclined to make most use of
- We automate and continuously improve request handling, to reduce time-to-value and irrational variation and the costs and effort that come with it
- We have a graceful method for rejecting service requests where the requestor is not entitled to what has been requested

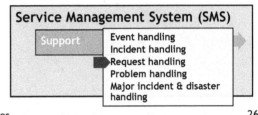

When rejecting a request for which the requestor is not entitled, we must always provide the reason to the requestor in a graceful way.

SMS - Support - Request handling - practices

M1
- ■ Request handling
- ■ Self-service portal

M2
- ■ Request handling
- ■ Multi-channel service
- ■ CAMS (for automation)

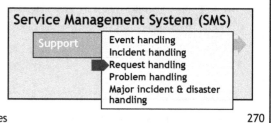

The key idea here is provide channels to stakeholders that stakeholder want to use, that make making requests "ready-to-hand", easy to get to, easy to do, just easy. And further, to automate, automate, automate, with self-service on the front-end, and scripted fulfillment on the back end.

SMS - Support - Problem handling - definition

Problem handling

Getting to the root cause of problems (that is, the unknown, underlying cause of one or more incidents) related to stakeholders, services, or the service management system, and raising a request for change to resolve them where it is justified to do so. Proactively preventing problems, and reactively, when a problem is live, working to minimize disruption the problem may cause on stakeholders, services, and the SMS; after a problem has occurred, reviewing the problem to identify action items for how to prevent similar problems from occurring, or to better handle similar problems should the occur in the future, and communicating the problem root cause, handling and action to stakeholders.

Problem

Root cause of one or more incidents NOTE The root cause is not usually known at the time a problem record is created and the problem management process is responsible for further investigation.
Source: ISO/IEC 20000-1:2011(E) 3 Terms and definitions

Known error

Problem that has an identified root cause or a method of reducing or eliminating its impact on a service by working around it. Source: ISO/IEC 20000-1:2011(E) 3 Terms and definitions

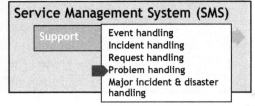

05-04-05. SMS – Support - Problem handling – Definition, desired state, best practices

A problem is the root cause, the unknown, underlying of one or more incidents. What capability do we have to handle problems?

Performance is effective when...

- At any give time, we can list the top 'n' problems we're working on, what we've done so far, and what we're going to do next

- We work proactively to prevent problems up front, and to prevent them from reoccurring; when a problem is live, we work to minimize the disruption the problem causes, through workarounds and crisp communication and action

- We have effective shared root cause analysis in "muscle memory", that multiply our capability to solve problems as a group

- Feedback to Learning & improvement for improving both services and the service management system to better handle similar situations going forward

- We practice blameless postmortems to review problems, which focus on learning and improvement, and are fact-driven and action-oriented, balancing accountability with a safe environment to share failures

- We arm problem solvers with the right skillset and authorization to handle problems

Service Management System (SMS)

Support	Event handling
	Incident handling
	Request handling
	▶Problem handling
	Major incident & disaster handling

05-04-05. SMS - Support - Problem handling - Definition, desired state, best practices

While traditional ITSM guidance describes problem handling as a process, and while you could draw a process map for it, there wouldn't be much to it. In the end, problem management is the sum of your application of shared techniques for preventing and solving problems, techniques like Kepner-Tregoe problem analysis.

SMS - Support - Problem handling - practices

M1

■ <u>Problem handling</u>

M2

■ <u>Problem handling</u>
■ <u>Site Reliability Engineering</u>
■ <u>Blameless postmortems</u>

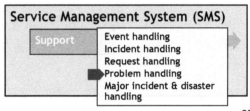

Service Management System (SMS)

Support	Event handling
	Incident handling
	Request handling
	Problem handling
	Major incident & disaster handling

Site reliability engineering (or SRE), is defined by Ben Treynor, founder of Google's Site Reliability Team, as "what happens when a software engineer is tasked with what used to be called operations." SRE is a discipline that incorporates aspects of software engineering and applies that to operations with the goal of creating ultra-scalable and highly reliable software systems. SRE focuses on engineering continuous operations at the point of customer consumption.

The largest impact of SRE on traditional ITSM-driven shops is that the SRE is an entirely new role in many shops. Part systems administrator, part second tier support and part developer, SREs ask questions, acquire new skills and knowledge, and embrace automation, and have the coding skills to solve problems and ensure the highest levels of reliability and availability. SREs are involved in the handling and management of late-night incidents, escalations, root cause analysis, service level objectives, availability and performance metrics, and other ITIL-driven practices. SREs provide a basis for instilling agility into many ITIL processes to meet changing demands brought about by the march to the cloud and DevOps, Lean and Agile practices.

Major incident & disaster handling

Fixing major incidents (that is, fixing something that is broken that is somewhere between a ordinary incident and a disaster), with a separate procedure with a tighter timeline and more urgency than a regular incident; and handling disasters, the abnormal situation where the business has been interrupted—you are literally "out of business", and you need to bring back services, in priority order, to get back into business. Both major incident and disaster handling have a proactive and a reactive aspect.

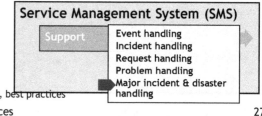

05-04-06. SMS - Support - Major incident & disaster handling - Definition, desired state, best practices

Module 5: Service Management Systems concepts, desired states and practices

274

You can think of a major incident as an incident that is much more significant than the average incident, that, if left untended, can become a disaster.

Traditional ITSM guidance places major incident management under incident management as a process, and refers to disaster handling as continuity management, which it places in a service design phase. OSM see major incidents and disasters as things worth managing, to a desired state. Note that some of the service qualities in the OSM model directly correspond to these two "things worth managing"—for example, supportability and recoverability. In OSM, these are not processes, but qualities built into services that must be kept in a desired state.

SMS - Support - Major incident & disaster handling - desired state

Performance is effective when...

- Major incidents and disasters are handled effectively and efficiently, within service level objectives or targets agreed with stakeholders
- Feedback to Learning & improvement for improving both services and the service management system to better handle similar situations going forward, and we've applied learning and automation to improve recoverability capability and speed by design up front and continual improvement
- We work proactively to prevent major incidents and disasters, and to prevent their reoccurrence; when a major incident or disaster is live, we work to minimize the disruption it causes, through workarounds, crisp communication and action
- We have effective shared major incident and disaster command systems in "muscle memory", that multiply our capability to solve problems as a group
- We practice blameless postmortems to review problems, which focus on learning and improvement, and are fact-driven and action-oriented, balancing accountability with a safe environment to share failures

Service Management System (SMS)	
Support	Event handling
	Incident handling
	Request handling
	Problem handling
	Major incident & disaster handling

05-04-06. SMS - Support - Major incident & disaster handling - Definition, desired state, best practices

Module 5: Service Management Systems concepts, desired states and practices

275

SMS - Support - Major incident & disaster handling - practices

M1

- Major incident & disaster handling
- Disaster handling

M2

- Public status pages / transparent uptime
- Incident Command System (ICS)

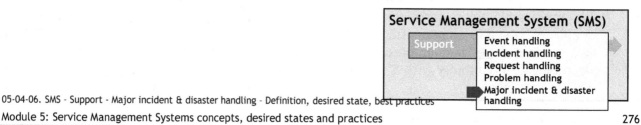

Service Management System (SMS)

Support	Event handling
	Incident handling
	Request handling
	Problem handling
	Major incident & disaster handling

Incident Command System is applicable to both major incidents and disasters, as it is a method of scaling up or down with self-similar teams.

Lesson summary

- Support is the practice area of the SMS that directs, controls and executes the day-to-day support of services, including handling events, requests, incidents, problems, major incidents and disasters. Contrast support with delivery, which is the component of the SMS that directs, controls and executes the day-to-day operation of services, including stakeholder relations, administration, provisioning, metering, and billing, and budgeting and accounting. Support is characterized by handling of customer and user requests within a timeline set in an SLA, as well as major incidents, problems, and disasters; while there are reactive aspects of the delivery components of the SMS, they are primarily characterized by planned activities on a set schedule, whereas support is primarily characterized by responding to customer and user requests, as well as problems, major incidents and disasters, with proactive aspects being important, but less frequent, than those that are unplanned.

- Each of these practice areas has a desired state that must be achieved and maintained continuously over time and through changing circumstances to ensure value continually flows to all stakeholders; these desired states can be achieved and maintained through best practices

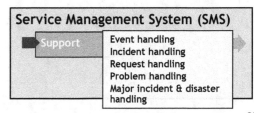

Delivery

05-05-01. SMS – Delivery
05-05-02. Stakeholder relations
05-05-03. Administration
05-05-04. Provisioning, metering & billing
05-05-05. Budgeting & accounting

Module 5 Lesson 4

SMS concepts, desired states and practices – Delivery (90m)

278

The purpose of this unit is to help you list and describe the what a service management system is, its typical components, their desired states, and best practices for achieving them, including the following SMS components in the category Delivery.

Delivery

The component of the SMS that directs, controls and executes the day-to-day operation of services, including stakeholder relations, administration, provisioning, metering, and billing, and budgeting and accounting. Contrast delivery with support, which is the component of the SMS that directs, controls and executes the day-to-day support of services, characterized by handling of customer and user requests within a timeline set in an SLA, as well as major incidents, problems, and disasters; while there are planned proactive aspects of the support components of the SMS, they are primarily characterized by responding to situations, whereas delivery is primarily characterized by planned activities executed on a set schedule, with reactive aspects being important, but less frequent, than those that are planned.

Service Management System (SMS)	
Delivery	Stakeholder relations Administration Provisioning, metering & billing Budgeting & accounting

Figure 05-05-01.1 SMS Delivery
05-05-01. SMS - Delivery - Definition, desired state, best practices

Module 5: Service Management Systems concepts, desired states and practices   279

Stakeholder relations is a "thing worth managing" in OSM, related to BRM in traditional ITSM guidance, but extending the idea to all stakeholders—customers, users, provider staff, and suppliers, and substituting a desired state (end-in-mind) for a process (means).

Administration consists of adding, modifying and deleting service subscriptions, add-ins and integrations, licenses, users, groups, resources, account and billing settings; getting sysadmin-level support for subscribed services; reviewing reports on usage, security and compliance; monitoring service health.

Provisioning, metering, and billing is concerned with initial set up of services for customers, monitoring their usage, and charging customers for the services they use. It also covers related items like handling upgrades, downgrade and cancellations that affect billing. In organizations where chargeback is not used, this area is concerned with showback.

Lastly, being able to budget and account for IT services is a "thing worth managing" that is part of the delivery of services.

You may see a departure here form some traditional ITSM guidance, that positions "delivery" as such things as service level, availability, capacity, continuity, and financial management. As you can see, OSM includes only financial management (budgeting & accounting) is service delivery. In OSM, availability, capacity (performance) and continuity (recoverability) are not processes, as in traditional ITSM, but qualities of IT-Led services.

SMS – Delivery – overall desired state

Performance is effective when...

- Stakeholders are satisfied with the delivery of our services, as time passes and stakeholder and overall industry circumstances change, as new technologies and possibilities emerge, and as our services change, including:
  - Stakeholders feel listened to, and feel we are adapting to new possibilities and changing circumstances appropriately and quickly, so that our services and their functionality and qualities reflect their feedback and needs
  - Stakeholders feel administrative transactions are quick and easy to administer, and that that there is an appropriate level of automation and self-service to them
  - Stakeholders feel that the process of provisioning services is quick and painless, that how services are metered is easy to understand up front and transparent, and that the billing or showback process is quick and easy and has the right options to meet their needs
  - Services are budgeted for accurately, so there are no unpleasant surprises like cost over-runs, and accounted for accurately, so there are no surprises as to costs, expenses, or profits
  - Stakeholders feel both planned and unplanned delivery activities are handled well

Service Management System (SMS)	
Delivery	Stakeholder relations Administration Provisioning, metering & billing Budgeting & accounting

These desired states are for the "things worth managing" in service delivery, which is the day-to-day administrative operations of services, including:

- Stakeholder Relations

- Administration

- Provisioning, metering & billing

- Budgeting & Accounting

SMS – Delivery – practices

M1	M2
Business Relationship Management	Stakeholder relations
Financial Management	■ Customer relations
	■ User relations
	■ Provider relations
	■ Supplier relations
	Administration
	Provisioning, metering, and billing
	Budgeting & accounting

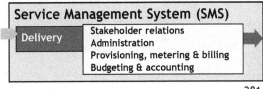

Best practices are what people do now that works to help them achieve and maintain desired states for things worth managing like stakeholders, day-to-day administration, provisioning, metering, and billing, and budgeting and accounting.

Stakeholder Relations

Engendering a constructive, productive and mutually beneficial relationship between the provider and customers, users, and suppliers, as well as between the provider as an organization and its staff; the desired relationship characterized by providing value in every interaction, and seeking to understand stakeholder circumstances and needs and technologies and the possibilities they create as they change, and to be understood as to the wherewithal required to consistently and sustainably provide value, and that results in the provider achieving and maintaining a continuous state where value flows between customers and users and the provider and suppliers, through a catalog of services that meets stakeholder needs as time passes and circumstances change, including stakeholder needs and what new technologies make possible.

Service Management System (SMS)

Delivery	Stakeholder relations
	Administration
	Provisioning, metering & billing
	Budgeting & accounting

05-05-02. SMS – Delivery – Stakeholder relations – Definition, desired state, best practices

You should see the difference here between stakeholder relations and BRM. BRM is a traditional ITSM process; the outcome of BRM is valid, and included in stakeholder relations in OSM, but extended to apply to all stakeholders, including provider staff.

Performance is effective when...

- Customers are satisfied both with the value of our services and with how we engage with them as time passes and circumstances change and as we adapt our services to those changes; our relationship with customers is constructive, productive and mutually beneficial, and characterized by our service portfolio continuously being adapted based on a combination of continuous customer listening and our bringing forward the possibilities new technologies and changing business circumstances bring

- Users are satisfied both with the use of our services and with how we engage with them as time passes and circumstances change and as we adapt our services to those changes; we have a constructive, productive and mutually beneficial relationship with users characterized by providing value in every interaction.

- Suppliers are satisfied with the value they get from our constructive, productive and mutually beneficial relationship; they consistently meet or exceed their commitments to us, and strive to bring more to the table than simply fulfilling their obligations

- We, the provider staff, are satisfied with our work, over time and through change

Service Management System (SMS)

Delivery	Stakeholder relations
	Administration
	Provisioning, metering & billing
	Budgeting & accounting

In the end, what we are aiming for is a sustainable situation where all key stakeholders are satisfied. This is a dynamic system, since what will satisfy stakeholders will change, as will the means, e.g., the technology possibilities. Note here that satisfying stakeholders is not meant to mean everyone is obliviously happy and gets everything they want, like a parent giving a child ice cream for breakfast, lunch and dinner. Relationships take work, sometimes hard, and educational conversations in both directions about what is needed, what is possible, and what wherewithal is required to deliver it consistently.

SMS – Delivery – Stakeholder relations – practices

Stakeholder relations
- <u>Customer relations</u>
- M1
    - <u>Business Relationship Management</u>
    - <u>Service Level Management</u>
- <u>User relations</u>
- M1
    - <u>Service Desk</u>
- <u>Provider relations</u>
- <u>Supplier relations</u>

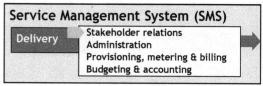

Service Management System (SMS)

| Delivery | Stakeholder relations
Administration
Provisioning, metering & billing
Budgeting & accounting |

In OSM, service level management is accomplished by making sure the service has the right configuration, features (including instrumentation), and qualities (which are the things like availability and performance (capacity) that go into a service level agreement, and making sure these "things worth managing" are kept in a desired state, over time and through changing circumstances.

SMS – Delivery - Administration – definition

Administration

The technical (sysadmin) component of adding, modifying and deleting service subscriptions, add-ins and integrations, licenses, users, groups, resources, account and billing settings; getting sysadmin-level support for subscribed services; reviewing reports on usage, security and compliance; monitoring service health.

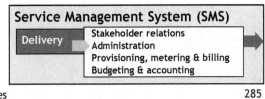

In delivery of services, someone's got to do the day to day sysadmin tasks, and they need to be done right—accurately, and in a timely manner.

SMS – Delivery – Administration – desired state

Performance is effective when:

- Stakeholders are pleased with responsiveness and quality of our conduct of technical (sysadmin) tasks associated with adding, modifying and deleting service subscriptions, add-ins and integrations, licenses, users, groups, resources, account and billing settings
- Stakeholders are pleased with the responsiveness and quality of our sysadmin-level support for subscribed services
- We review reports on usage, security and compliance, and monitoring service health, to avoid issues and outages
- Wherever possible and cost-effective, we automate sysadmin tasks to reduce time-to-value, as well as unnecessary variation and its associated costs

Service Management System (SMS)	
Delivery	Stakeholder relations
	Administration
	Provisioning, metering & billing
	Budgeting & accounting

Automating sysadmin tasks is a key part of this, and the SRE role can be a key part of this.

SMS – Delivery – Administration – practices

M1

- Sysadmin

M2

- Site Reliability Engineering

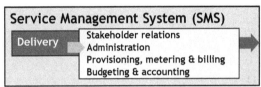

Service Management System (SMS)

Delivery	Stakeholder relations
	Administration
	Provisioning, metering & billing
	Budgeting & accounting

Site Reliability Engineering, which is described as "what happens when a software engineer is tasked with what used to be called operations", is a discipline that incorporates aspects of software engineering and applies them to IT operations problems, with the aim of creating ultra-scalable and highly reliable software systems.

Provisioning, metering & billing

Setting up (and later, decommissioning) services and service components; tracking and reporting on usage; invoicing customers for services, either with chargeback or showback.

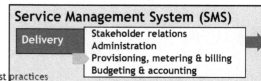

Service Management System (SMS)

Delivery	Stakeholder relations
	Administration
	Provisioning, metering & billing
	Budgeting & accounting

Provisioning, metering and billing is the set of activities required to bill customers for services they consume (or, if you don't do chargeback, at least showback). It requires sound IT accounting practices and systems, and includes a periodic negotiation cycle to set (usually annually) pricing policy and price lists, monthly compiling and issuing of bills (either charge- or show-back).

Performance is effective when...

- Stakeholders are satisfied both with the efficiency and results of provisioning (including initial setup and later, decommissioning) as well as with the engagement / interaction itself

- Stakeholders are satisfied with the accuracy of metering and billing (whether it is chargeback or showback), including the transparency of algorithms for both, and the provisions we make to allow them to influence their usage and costs to their advantage

- If we charge back, our charging model is transparent and communicated up front so there are no negative surprises to customers, and it is easy for customers to pay and for us to collect payment; we are also clear on our goal for charging (e.g., loss leader, break even, profit) and our algorithms

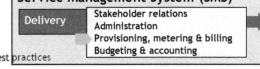

05-05-04. SMS – Delivery – Provisioning, metering & billing – Definition, desired state, best practices

Chargeback is providing pay-the-bills customers with an analysis of the provider service costs on, for example, a monthly basis, and actually charging them for those costs. Showback (also known as memo billing), is providing the same analysis, but without actually charging them for those costs. The idea with showback is to provide information to drive the right behaviors. Showback is similar to the note you get from your insurance provider for a medical procedure you have had that is fully covered—this is what it would have cost you, etc.

Some other indicators of performance include:

- Customers can easily determine what services have and will costs, and can easily manage their subscription through a portal

- Stakeholders see Total Cost of Ownership (TCO) for services as reasonable

SMS – Delivery – Provisioning, metering & billing – practices

M1

■ <u>Financial Management</u>

M2

Provisioning, metering, and billing

■ <u>Consumptive billing model</u>

■ <u>Chargeback and showback</u>

■ <u>Provisioning (User)</u>

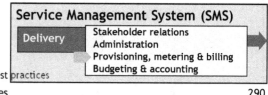

Service Management System (SMS)

Delivery	Stakeholder relations
	Administration
	Provisioning, metering & billing
	Budgeting & accounting

We can be doing well as a provider in a lot of areas, but screwing up on "the last mile"—on things that touch the client directly and greatly impact their preferences and perceptions—is something we cannot afford to do. Provisioning, metering and billing are part of that last mile.

Budgeting & Accounting

Budgeting is the process of preparing a budget, which is a set of interlinked plans that quantitatively describe an entity's projected future operations. A budget is used as a yardstick against which to measure actual operating results, for the allocation of funding, and as a plan for future operations.

Accounting is the systematic recording of the financial transactions of a business. Such recording can be split into three activities: 1) Setting up a system of record keeping, 2) Tracking transactions within that system of record keeping, and 3) Aggregating the resulting information into a set of financial reports

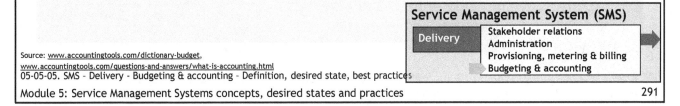

Source: www.accountingtools.com/dictionary-budget,
www.accountingtools.com/questions-and-answers/what-is-accounting.html
05-05-05. SMS – Delivery – Budgeting & accounting – Definition, desired state, best practices

Module 5: Service Management Systems concepts, desired states and practices

291

It is important that budgeting and accounting systems be set up with a chart of accounts, reports, etc. that can record and show, e.g., budget versus actual, by services.

SMS - Delivery - Budgeting & accounting - desired state

Performance is effective when...

- We accurately account for IT costs, map cost data back to categories and services, and use that data for investment and budgeting decisions.

- Whether we offer services to make a profit or offer services at no profit, our services and the SMS have a sustainable funding model

- We can easily determine and communicates the full cost of providing services and the SMS and revenue and profit, and we are proactively managing costs down and revenue and profit and value to stakeholders up

- We can quantify the value of each service we provide, and of the SMS

- We operate in line with our organization's financial policies

- Expenses are the same or less, and revenues and profits are the same or higher than projected for services

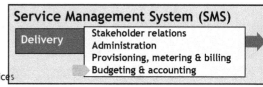

05-05-05. SMS - Delivery - Budgeting & accounting - Definition, desired state, best practices

Module 5: Service Management Systems concepts, desired states and practices

292

ISO/IEC 20000-1:2011(E) 6.4 Budgeting and accounting for services indicates that there must be policies and documented procedures for

a) budgeting and accounting for service components including at least

1) assets — including licenses — used to provide the services,

2) shared resources,

3) overheads,

4) capital and operating expenses,

5) externally supplied services,

6) personnel, and

7) facilities

b) apportioning indirect costs and allocating direct costs to services, to provide an overall cost for each service

c) effective financial control and approval.

SMS – Delivery - Budgeting & accounting - practices

■ <u>Budgeting & accounting</u>

Service Management System (SMS)

Delivery
- Stakeholder relations
- Administration
- Provisioning, metering & billing
- Budgeting & accounting

One key best practice here is to set up your chart of accounts and reports to be able to budget, account, and report on the planned and actual expenses and revenues, by service.

Lesson summary

- Delivery is the practice area within the SMS that covers:
    - Stakeholder relations – ensuring the relationship between the provider and customers, users and suppliers is good
    - Administration – ensuring administration of services is easy for all stakeholders
    - Provisioning, metering, and billing—setting up (and later, decommissioning) services and service components; tracking and reporting on usage; invoicing customers.
    - Budgeting & Accounting—preparing budgets for services and accounting for expense and revenue
- Each of these practice areas has a desired state that must be achieved and maintained continuously over time and through changing circumstances to ensure value continually flows to all stakeholders; these desired states can be achieved and maintained through best practices

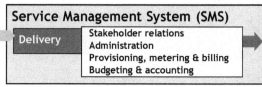

Module 6
Summary and exam preparation (120m)

At the end of this module, you should be able to:

- Summarize key points for each topic in the OSM Foundation course
- Identify topic areas to focus on and review to pass the exam
- Prepare yourself to take the OSM Foundation examination

OSM Foundation Course Agenda

Module 1: Service management and Open Service Management	60m
Module 2: Stakeholder concepts, desired states and practices	60m
Module 3: Value and value flow concepts, desired states and practices	60m
Module 4: Service concepts, desired states and practices	240m
Module 5: SMS concepts, desired states and practices	300m
Module 6: Summary and exam preparation	120m

The minimum contact hours for this course are 14, including 2 hours for sample examination revision.

In this module, we review what you've learned and take and review a sample exam to prepare you to take the OSM Foundation certification examination.

OSM Foundation Course Objectives

At the end of the course, you should be able to list and describe the core principles, practices, and terminology associated with service management, including:

- Service management and Open Service Management
- Stakeholder concepts, desired states and practices
- Value and value flow concepts, desired states and practices
- Service concepts, desired states and practices, including service configuration, functionality and qualities
- Service Management System (SMS) concepts, desired states and practices, including design and transition, promotion, support and delivery

Format of the Examination

Type	Multiple choice, 40 questions. The questions are selected from the full OSM Foundation Certificate in IT Service Management examination question bank.
Duration	60 minutes for all candidates in where the exam is in their native tongue; 75 minutes where it is not the native tongue.
Prerequisites	Training from an Registered Trainer (RT) from a Registered Training Organization (RTO).
Format	Supervised, closed book; available online or on paper.
Passing Score	26 of 40 or 65%

You must achieve a passing score in the Foundation Exam to gain the OSM Foundation Certificate in IT Service Management.

Tips for Taking the OSM Foundation Exam

- Attempt all 40 questions; if you don't know the answer, guess and mark for review
- There are no trick questions or patterns in how the answers are laid out
- Mark all answers on the paper answer sheet (or if taken online, on the web page)
- Skip and go back to a question you can't answer right away
- Don't forget to use the process of elimination
- Skip the preamble; it might be a misdirection, go right to the question; only read the preamble if it is necessary to answer the question
- Watch for absolutes, for example, "always", and "guarantees"
- If stuck between choices ask, "What is this, and who owns it?"
- Read questions carefully—don't miss the word "NOT"!
- For testing purposes, take the book view, rather than your own

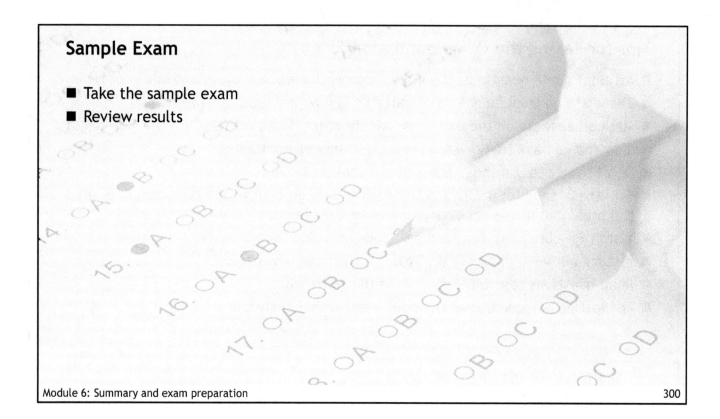

Sample Exam

- Take the sample exam
- Review results

Top Takeaways from This Course

- What are the top three things you're going to take back to your job?
- What are the top three things you will do when you get back?

This course will not be of value to you if you do not make a plan to understand further what OSM is all about.

What have you learned that you would like to share with others in your organization?

List down some of your personal commitments in terms of how you will apply what you have learned when you go back to work.

Next Steps

- Apply essential OSM concepts in your everyday work
- Learn more about IT Service Management and OSM
    - Visit osmalliance.org for more resources, updates
        - ♦ Industry presentations and webinars
        - ♦ Whitepapers, articles
        - ♦ Consulting and training services
        - ♦ Links to key resources
- Contribute to OSM as and individual, team, or organization
    - Sign up today at osmalliance.org
    - Contact us at info@osmalliance.org

Open Service Management®
OPEN COMMUNITY-DRIVEN BEST PRACTICE

Professional Qualifications for Open Service Management®

The Open Service Management Foundation Sample Exam A
Version 1.2, January 1, 2018

Instructions

1. *The examination is 40 questions, multiple choice. Attempt all questions.*

2. *26 or more correct answers out of 40 (65%) are needed to pass the exam.*

3. *Mark all answers on the provided answer sheet.*

4. *You have 60 minutes to complete the examination.*

1. Which of the following are NOT included in the list of things worth managing when it comes to service management:

a) Stakeholders

b) Services

c) Service Management Systems

d) Vendors

2. Which is NOT a type of best practice?

a) Deterministic

b) Vital

c) Adaptive

d) Emergent

3. When it comes to service, which is true for the customer?

a) Owns costs and risks, but not the details

b) Gets paid to perform services

c) Responsible for means of achieving desired states

d) Takes on the details of costs and risks

4. People or organizations that manage and deliver services to one or more internal or external customers are:

a) Users

b) Suppliers

c) Providers

d) Partners

5. Which of the following is NOT a type of stakeholder user:

a) End users

b) Lead users

c) Super users

d) Admin users

6. "Ensuring that all contracts support business needs" is a desired state for:

a) Suppliers

b) Users

c) Providers

d) Partners

7. How the worth, as assigned by the customer of a product or service, moves is known as:

a) Takt time

b) Value flow

c) Pace

d) Pull

8. Best practices for value flow include: 1) Value stream mapping, 2) Continuous integration, 3) Scheduled delivery

a) 1 & 2 only

b) 1 & 3 only

c) 2 & 3 only

d) All of the above

9. Which of the following is NOT true about the nature of value?

a) Value is in the mind of stakeholders

b) Only customers get to define value

c) Value is continuously changing

d) Value is bound to time and cost

10. The willingness to help customers and to provide prompt services is:

a) Responsiveness

b) Empathy

c) Assurance

d) Reliability

11. Examples of IT-Led Services are: 1) Azure, 2) Consulting, 3) Office365

a) 1 & 2 only

b) 2 & 3 only

c) 1 & 3 only

d) All of the above

12. Which is the BEST description of the desired state for data service configuration?

a) Performance is effective when it compares well across criteria categories such as Cloud Characteristics, Production Ready and DevOps and Lifecycle Phases.

b) Performance is effective when there is automation.

c) Performance is effective when there is automation, security requirements are met and feature and performance requirements are met.

d) Performance is effective when we know what software we have, it is being used and its functionality and service qualities meet what is required by customers and users.

13. Fill in the blank: Runtime is the runtime environment (RTE) within which the _____ component of a service executes, e.g., Java RTE.

a) Software

b) Data

c) Application

d) Platform

14. Computer operating systems perform basic tasks such as: 1) recognizing input from the keyboard, 2) keeping track of files and directories on the disk, 3) ensuring only authorized users access the system.

a) 1 & 2 only

b) 1 & 3 only

c) 2 & 3 only

d) All of the above

15. Which of the following is NOT a desired state for virtualization?

a) We backup early, backup often.

b) We centralize our storage.

c) We are careful when using "undo" technology.

d) We know who owns the infrastructure we are paying for.

16. When we leverage tiered storage, consolidate storage pools and implement staged backup to disk, we are aligned with the desired state for:

a) Server

b) Storage

c) Middleware

d) Data

17. Monitors, hard drives, memory and CPU are examples of:

a) Software

b) Storage

c) Platform

d) Hardware

18. Service functionality best practices include: 1) Stakeholders actively participate, 2) Inclusive Models are adopted, 3) Keep it fun

a) 1 & 2 only

b) 2 & 3 only

c) 1 & 3 only

d) All of the above

19. When it comes to Human-led service qualities, performance is effective when stakeholders rate services highly among the following dimensions:

a) Tangibles, Assurance, Regulation and Sympathy

b) Actuals, Regulations, Intangibles, Accuracy

c) Regulations, Responsiveness, Accordance and Accuracy

d) Tangibles, Responsiveness, Assurance and Empathy

20. The ability to gather information about the state of something and to control it is:

a) Manageability

b) Reliability

c) Availability

d) Responsiveness

21. The desired state of ensuring IT capacity matches business needs at the right time and at the right cost is related to:

a) Serviceability

b) Availability

c) Assurance

d) Performance

22. When we can continuously demonstrate that data has not been corrupted, we are aligned with the desired state of:

a) Reliability

b) Integrity

c) Assurance

d) Responsiveness

23. The ability to communicate, exchange data and work with other services is:

a) Compliability

b) Interoperability

c) Internationalization

d) Responsiveness

24. Which of the following is NOT an aspect of the overall desired state of a Service Management System?

a) How services are planned

b) How services are developed

c) How services are measured

d) How services are bought and sold

25. Design & transition are the set of SMS practices areas that are concerned with: 1) Alignment, Learning & Improvement of services, 2) Development of services, 3) Retirement of Services

a) 1 & 2 only

b) 2 & 3 only

c) 1 & 3 only

d) All of the above

26. Non-fulfilment of a requirement is known as:

a) Continual improvement

b) Corrective action

c) Nonconformity

d) Preventative action

27. Which of the following is NOT a development best practice?

a) Testing

b) User Stories

c) Continuous Delivery

d) Application Lifecycle Management

28. Planning, scheduling and controlling the build, test and deployment of releases with minimal disruption is known as:

a) Release & deployment

b) Change handling

c) Continuous delivery

d) Corrective action

29. Policies, standards, models, templates, workflows and samples describe:

a) Best practices

b) Standardized approaches

c) User manuals

d) Service Design Package (SDP)

30. SMS component change handling best practices include: 1) Complexity, variation and dependency reduction, 2) Risk management, 3) Evolved implementation.

a) 1 & 2 only

b) 2 & 3 only

c) 1 & 3 only

d) All of the above

31. A collection of one or more new or changed configuration items released into the live environment is known as a:

a) Deployment

b) Change

c) Retirement

d) None of the above

32. The marketing, advertising and selling of existing external services and new and changing external services is known as:

a) Design & Transition

b) Promotion

c) Support

d) Delivery

33. Stakeholders indicating that events, requests and incidents are handled within established timelines is a desired state for:

a) SMS Support

b) SMS Promotion

c) SMS Design & Transition

d) SMS Delivery

34. Monitoring for and detecting changes in state of significance for stakeholders, services or SMS components is:

a) Request handling

b) Problem handling

c) Event handling

d) Incident handling

35. Best practices for event handling include: 1) Automation, 2) SLAM charts, 3) Monitoring and logging

a) 1 & 2 only

b) 2 & 3 only

c) 1 & 3 only

d) All of the above

36. Unplanned interruptions to a service and reductions in the quality of service are examples of:

a) Events

b) Incidents

c) Problems

d) Requests

37. Incident handling best practices include: 1) Standardized methods and procedures, 2) Service Desk and ticketing system, 3) risk assessment and management

a) 1 & 2 only

b) 2 & 3 only

c) 1 & 3 only

d) All of the above

38. A problem that has an identified root cause or a method of reducing or eliminating its impact on a service by working around it is a:

a) Known error

b) Unknown error

c) Incident

d) Problem

39. Examples of SMS Administration best practices are: 1) Risk Management, 2) Retirement Planning, 3) Service Level Management

a) 1 & 2 only

b) 2 & 3 only

c) 1 & 3 only

d) All of the above

40. Performance is effective for an SMS component configuration when 1) we ensure a logical model of IT services, 2) we know what configuration items we have out there, 3) we can produce accurate information to meet audit and regulatory requirements

a) 1 & 2 only

b) 2 & 3 only

c) 1 & 3 only

d) All of the above

Open Service Management®
OPEN COMMUNITY-DRIVEN BEST PRACTICE

Professional Qualifications for Open Service Management®

The Open Service Management Foundation Sample Exam A
Answers and Rationale, Version 1.2, January 1, 2018

Q	Reference	A	Explanation
1	01-01	D	Stakeholders, value flow, services, and the service management system are all "things worth managing". Suppliers are a type of stakeholder.
2	01-03	B	The three types of practices are deterministic, adaptive, and emergent.
3	01-05	A	Customers own the costs and risks of services, but the details of those costs and risks are handled by the provider.
4	02-01	C	Service providers provide services.
5	02-03	B	End, Super, and Admin users are all valid types of users.
6	02-05	A	While all stakeholders may benefit from contracts that support business needs, ensuring that they do is a desired state for suppliers.
7	03-01	B	Value flow is how values moves; takt time is a measure from the start of production of one unit and the start of production of the next.
8	03-01	A	Value stream mapping and continuous integration both can contribute to value flow.
9	03-02	B	Customers seeing value in services is key, but value must flow to all stakeholders of a service for the service to be sustainable.
10	04-01-03	A	For human-led services, being responsive is a function of how capable and willing you are to provide a service, and how quickly you do it.
11	04-01-04	C	Azure and Office365 are examples of services that are mostly automated, with humans on the back end, and as such they are IT-Led. Consulting services typically are provided by humans, with varying degrees of support and involvement by technology, thus they are human-led.
12	04-01-07	B	Book answer.
13	04-01-09	C	Book answer.
14	04-01-11	D	Book answer.
15	04-01-13	D	Backing up early and often, centralizing storage, and being careful when using "undo" technology are all desired states for virtualization.

#	Code	Ans	Explanation
16	04-01-15	B	These are all desired states for storage.
17	04-01-17	D	Hardware are physical elements of your infrastructure.
18	04-02-01	D	When deciding functional requirements, inclusion and active participation by all key stakeholders, as well as keeping it fun, are key.
19	04-02-03	D	We as consumers judge the services we consume based on tangibles (for example, what a physical store might look like), responsiveness (how quickly did we get served), our trust in the capabilities of those providing the services, and their desire to help us through their service.
20	04-02-11	A	A thing is manageable to the extent we can understand and influence its state.
21	04-02-13	D	Performance suffers when capacity does not meet demand, resulting in, for example, slow transaction turnaround time.
22	04-02-19	B	Data has integrity to the extent it has not been corrupted in some way.
23	04-02-27	B	A system has interoperability to the extent other systems can interact effectively with it, for example, through a command line interface or API.
24	05-01-01	D	Your service management system is your capability to manage services in all phases, including planning, development, and measurement.
25	05-02-01	D	Design & transition practices are about adding, changing, and removing services.
26	05-02-03	C	Book answer.
27	05-02-04	C	CD is listed under Release & Deploy practices in OSM.
28	05-02-05	A	Book answer.
29	05-02-05	B	One way to coordinate work and minimize variation and waste is through policies, standards, models, templates, workflows and samples.
30	05-02-05	A	Book answer.

#	Code	Ans	Explanation
31	05-02-05	A	Book answer.
32	05-03-01	B	Promotion consists of internal or external marketing and "selling" activities, including activities to ensure uptake of a service in the first place, and uptake on features as they are introduced to ensure value.
33	05-04-01	A	Handling events, requests, and incidents falls under Support.
34	05-04-02	C	Events are changes in state of significance, and can be informational, warnings (as when a pre-set threshold has been reached) or exception such as a service being down.
35	05-04-02	C	SLAM charts are related to service level and availability management.
36	05-04-03	B	Book answer.
37	05-04-03	A	Having standardized methods and procedures, a service desk, and a ticketing system are all best practices for incident handling.
38	05-04-05	A	A problem is an unknown underlying cause; when the cause is determined, it is a known error.
39	05-05-03	A	Service level management falls under stakeholder relations
40	05-02-06	D	All three are required for a sufficient logical picture of our environment.

Open Service Management®
OPEN COMMUNITY-DRIVEN BEST PRACTICE

Professional Qualifications for Open Service Management®

The Open Service Management Foundation Sample Exam B
Version 1.2, January 1, 2018

Instructions

1. *The examination is 40 questions, multiple choice. Attempt all questions.*

2. *26 or more correct answers out of 40 (65%) are needed to pass the exam.*

3. *Mark all answers on the provided answer sheet.*

4. *You have 60 minutes to complete the examination.*

1. What are "desired states"?
   a) Outputs

   b) Deliverables

   c) Outcomes

   d) The "end in mind" for things worth managing

2. What are the four primary stakeholders of service management?

   a) Customers, Users, Providers, Suppliers

   b) Consumers, Providers, Managers, Systems

   c) Users, Boards, Suppliers, Partners

   d) Suppliers, Customers, Leaders, Vendors

3. Which is NOT a component of value?

   a) Price

   b) Functionality

   c) Deficiencies

   d) Qualities of the service

4. People or organizations that are third parties who supply goods or services are:

   a) Partners

   b) Suppliers

   c) Users

   d) Providers

5. Managers, developers and infrastructure engineers are types of roles in:

   a) Users

   b) Partners

   c) Providers

   d) Suppliers

6. "Contract review, renewal, termination" is a best practice for:

    a) Suppliers

    b) Users

    c) Providers

    d) Partners

7. Which of the following is NOT a desired state for value flow?

    a) Value flows to all stakeholders unencumbered

    b) Value flow is visible and managed

    c) Value flows to customers only

    d) Time-to-value is acceptable for all stakeholders

8. Components of value include: 1) price, 2) functionality, 3) qualities of the service

    a) 1 & 2 only

    b) 1 & 3 only

    c) 2 & 3 only

    d) All of the above

9. The Agile Manifesto values Individuals and Interactions over:

    a) Following a plan

    b) Contract negotiation

    c) Processes and tools

    d) Working software

10. RACI and SERVQUAL are best practices related to:

    a) The Deming Cycle

    b) The Pareto Principle

    c) IT-Led Services

    d) Human-Led Services

11. Which is the BEST description of the desired state for software service configuration?

   a) Performance is effective when we know what software we have, it is being used and its functionality and service qualities meet what is required by customers and users.

   b) Performance is effective when there is automation, security requirements are met and feature and performance requirements are met.

   c) Performance is effective when there is automation.

   d) Performance is effective when it compares well across criteria categories such as Cloud Characteristics, Production Ready and DevOps and Lifecycle Phases.

12. Which is the BEST description of the desired state for platform service configuration?

   a) Performance is effective when there is automation, security requirements are met and feature and performance requirements are met.

   b) Performance is effective when there is automation.

   c) Performance is effective when we know what software we have, it is being used and its functionality and service qualities meet what is required by customers and users.

   d) Performance is effective when it compares well across criteria categories such as Cloud Characteristics, Production Ready and DevOps and Lifecycle Phases.

13. Software that serves to "glue together" separate, often complex and already exisiting programs is known as:

   a) Micro-services architecture

   b) Middleware

   c) Containers

   d) Object Linking and Embedding

14. Which of the following is NOT a desired state for infrastructure?

   a) We have the level of access we need.

   b) We have ensured that different programs and users running at the same time do not interfere with eachother.

   c) We know how much modification our software will require.

   d) We have ensured compliance.

15. A computer, device or program dedicated to managing network resources is a:

    a) Application

    b) Software

    c) Server

    d) Platform

16. A group of two or more devices that can communicate is a(n):

    a) Network

    b) Server

    c) Platform

    d) Application

17. The physical service environment is known as:

    a) Storage

    b) Network

    c) Platform

    d) Facilities

18. Which of the following best describes the definition of service qualities?

    a) What a service is made of

    b) How a service behaves

    c) What a service does

    d) How a service functions

19. The ability to demonstrate willingness to help customers and provide prompt service is:

    a) Reliability

    b) Availability

    c) Assurance

    d) Responsiveness

20. The ability to identify issues and take corrective action such as to repair or upgrade a component in a running system is:

   a) Manageability

   b) Serviceability

   c) Reliability

   d) Responsiveness

21. The ability to protect information from unauthorized access source is known as:

   a) Manageability

   b) Serviceability

   c) Security

   d) Responsiveness

22. The desired state for _____, is when services conform to all applicable legislative and regulatory requirements.

   a) Reliability

   b) Integrity

   c) Assurance

   d) Compliability

23. When services (and the SMS) can adapt automatically to workload changes by provisioning and de-provisioning resources in an automatic manner, we are aligned with the desired state for:

   a) Compliability

   b) Interoperability

   c) Elasticity

   d) Portability

24. The support component of an SMS directs, controls and executes the day-to-day support of services including handling: 1) events & requests, 2) invoicing & costs, 3) problems & major incidents

a) 1 & 2 only

b) 2 & 3 only

c) 1 & 3 only

d) All of the above

25. Which of the following is NOT a best practice of SMS design & transition?

a) Service Design Package (SDP)

b) Value Stream Mapping

c) Continuous Integration

d) Continuous Delivery

26. An action taken to eliminate the cause or reduce the likelihood of recurrence of a detected nonconformity or other undesireable situation is known as:

a) Continual improvement

b) Corrective action

c) Nonconformity

d) Preventative action

27. Planning, building and testing new stakeholder satisficing mechanisms, services and SMS components as well as making improvements to all three is known as:

a) Development

b) Deployment

c) Implementation

d) Integration

28. Coordinating changes related to stakeholders, services, or the SMS within timelines set by service level targets (SLTs) describes:

    a) Release & deployment

    b) Corrective action

    c) Change handling

    d) Course correction

29. Scheduling, classification, prioritization, variation and dependencies are all aspects of:

    a) Delivery

    b) Standardized approaches

    c) Lifecycle support

    d) Automation

30. A change model is a repeatable workflow for handling a particular category of change meant to:

    a) Increase speed

    b) Reduce risk

    c) Reduce unnecessary variation

    d) All of the above

31. Best practices in _____ are Marketing BP, Selling BP and SEO.

    a) Promotion

    b) Design & Transition

    c) Delivery

    d) Support

32. Desired states for event handling include: 1) Feedback to Learning & Improvement to better handle situations, 2) Regular monitoring and reporing on stakeholder satisfaction, 3) Defined reporting and review mechanism

    a) 1 & 2 only

    b) 2 & 3 only

    c) 1 & 3 only

    d) All of the above

33. Instrumentation to trap something that is broken so that it can be fixed within timeframes set by service level targets (SLTs) is:

    a) Event handling SMS telemetry

    b) Incident handling SMS telemetry

    c) Request handling SMS telemetry

    d) Administration SMS telemetry

34. When feedback to Learning & Improvement for improving both services and the service management system to better handle similar situations going forward, performance is effective for:

    a) Request handling

    b) Problem handling

    c) Event handling

    d) Incident handling

35. Best practices for Problem handling include: 1) Service Desk and ticketing system, 2) Blameless post-mortems, 3) Root cause analysis techniques

    a) 1 & 2 only

    b) 2 & 3 only

    c) 1 & 3 only

    d) All of the above

36. The root cause of one or more incidents is a:

    a) Known error

    b) Unknown error

    c) Incident

    d) Problem

37. Handling _____ is respondng to someone asking for something they are entitled to - unlike incidents, where something needs to be fixed.

    a) Problems

    b) Events

    c) Requests

    d) Major Incidents

38. Which statement does NOT reflect the delivery component of the SMS?

    a) Directs, controls & executes day-to-day operations

    b) Directs stakeholder relations

    c) Unplanned and reactive activities

    d) Controls provisioning

39. Adding, modifying and deleting service subscriptions, add-ins and integrations are examples of:

    a) Provisioning

    b) Budgeting

    c) Administration

    d) Billing

40. Setting up and later decommissioning services and service components, tracking and reporting on usage, and invoicing customers for services are known as:

    a) Deploying, Billing, Metering

    b) Metering, Implementing, Administering

    c) Provisioning, Administering, Billing

    d) Provisioning, Metering, Billing

Open Service Management®
OPEN COMMUNITY-DRIVEN BEST PRACTICE

Professional Qualifications for Open Service Management®

The Open Service Management Foundation Sample Exam B
Answers and Rationale, Version 1.2, January 1, 2018

Q	Reference	A	Explanation
1	01-02	D	Book answer.
2	01-05	A	Book answer.
3	01-06	C	Components of value include price, functionality (features) and qualities (things like availability, accessibility and the like).
4	02-01	B	Book answer.
5	02-04	C	Book answer.
6	02-05	A	Book answer.
7	03-01	C	Value must flow to all stakeholders, including customers, users, the provider and its suppliers, for a service to be sustainable.
8	03-01	D	Book answer.
9	03-05	C	See agilemanifesto.org
10	04-01-03	D	RACI is a technique for understanding ownership and involvement levels of roles relative to tasks in workflows, and SERVQUAL is a service quality model commonly applied to professional or human-led (as opposed to IT-led) services.
11	04-01-05	A	Book answer.
12	04-01-08	D	Book answer.
13	04-01-10	B	Middleware is software that acts as a bridge or connector between services and applications.
14	04-01-12	B	Answer B is a task of an Operating System.
15	04-01-14	C	Book answer.
16	04-01-16	A	Book answer.
17	04-01-18	D	Facilities are the "bricks and mortar" environment of computing, such as an on-premise data center, with concerns like power and air conditioning, air flow and so on.
18	04-02-02	B	What a service is made of is its components; how a service behaves is its qualities; what a service does is its functionality.
19	04-02-05	D	This is a characteristic of human-led services.
2	04-02-12	B	Manageability is a function of the visibility you have into the health of the thing worth managing, along

#	Code	Answer	Explanation
0			with the ease (or difficulty) in taking action aimed at controlling it, for example, to restore it to a desired state.
21	04-02-18	C	Book answer.
22	04-02-22	D	Compliability is how compliant the system is with relevant regulatory requirements; Assurance is a characteristic of human-led services, and about establishing trust through effective performance and service orientation.
23	04-02-29	C	A system is elastic to the degree that it can adapt resource capacity to changes in demand by scaling resources up and down automatically, such that at any given time capacity matches demand.
24	05-01-01	C	Invoicing & costs are part of administration.
25	05-02-01	B	Value stream mapping is a best practice for value flow.
26	05-02-03	B	Book answer.
27	05-02-04	A	The scope of development is larger than simply application development, and extends to include anything required to deliver a service.
28	05-02-05	C	Change handling makes sure changes that are needed get done in a timely manner, without undue disruption, and that "what changed?" can be tracked.
29	05-02-05	D	Automation can support the first three, and can be used to reduce variation and dependencies, for example, through infrastructure automation.
30	05-02-05	D	Book answer.
31	05-03-01	A	Promotion includes marketing and selling (or in the case of internal services, promoting uptake of services).
32	05-04-02	D	Book answer.
33	05-04-03	B	Incident management is about fixing what is broken.
34	05-04-03	D	Book answer.
35	05-04-05	B	Book answer.
36	05-04-05	D	Problems are the unknown, underlying cause of one or more incidents.

3 7	05-04-06	C	Valid requests come from requestors who are entitled to what they are requesting as part of their subscription to a service.
3 8	05-05-01	C	Delivery is the supplying of the service--unplanned and reactive active fall under support.
3 9	05-05-03	C	Administration is the business operations end of running a service provider.
4 0	05-05-04	D	Provisioning is the act of setting up service, metering is tracking consumption, and billing is charging for that consumption.